Thermodynamic Formalism

The Mathematical Structures of Equilibrium Statistical Mechanics

Second Edition

Reissued in the Cambridge Mathematical Library this classic book outlines the theory of thermodynamic formalism which was developed to describe the properties of certain physical systems consisting of a large number of subunits. It is aimed at mathematicians interested in ergodic theory, topological dynamics, constructive quantum field theory, and the study of certain differentiable dynamical systems, notably Anosov diffeomorphisms and flows. It is also of interest to theoretical physicists concerned with the computational basis of equilibrium statistical mechanics. The level of the presentation is generally advanced, the objective being to provide an efficient research tool and a text for use in graduate teaching. Background material on physics has been collected in appendices to help the reader. Extra material is given in the form of updates of problems that were open at the original time of writing and as a new preface specially written for this edition by the author.

DAVID RUELLE is a Professor Emeritus at the Institut des Hautes Etudes Scientifiques, Bures-sur-Yvette, Paris.

Other books available in the Cambridge Mathematical Library

G. E. Andrews	*The Theory of Partitions*
H. F. Baker	*Abelian Functions*
A. Baker	*Transcendental Number Theory*
G. K. Batchelor	*An Introduction to Fluid Dynamics*
N. Biggs	*Algebraic Graph Theory, 2nd Edition*
J. C. Burkill, H. Burkill	*A Second Course in Mathematical Analysis*
S. Chapman, T. G. Cowling	*The Mathematical Theory of Non-uniform Gases*
R. Dedekind	*Theory of Algebraic Integers*
P. G. Drazin, W. H. Reid	*Hydrodynamic Stability, 2nd Edition*
G. H. Hardy	*A Course of Pure Mathematics*
G. H. Hardy, J. E. Littlewood, G. Pólya	*Inequalities, 2nd Edition*
D. Hilbert	*Theory of Algebraic Invariants*
W. V. D. Hodge, D. Pedoe	*Methods of Algebraic Geometry, Volumes I, II & III*
R. W. H. Hudson	*Kummer's Quartic Surface*
A. E. Ingham	*The Distribution of Prime Numbers*
H. Jeffreys, Bertha Jeffreys	*Methods of Mathematical Physics, 3rd Edition*
Y. Katznelson	*An Introduction to Harmonic Analysis, 3rd Edition*
H. Lamb	*Hydrodynamics, 6th Edition*
J. Lighthill	*Waves in Fluids*
M. Lothaire	*Combinatorics on Words, 2nd Edition*
F. S. Macaulay	*The Algebraic Theory of Modular Systems*
C. A. Rogers	*Hausdorff Measures, 2nd Edition*
L. C. G. Rogers, D. Williams	*Diffusions, Markov Processes and Martingales, 2nd Edition, Volumes I & II*
L. Santalo	*Integral Geometry and Geometric Probability*
W. T. Tutte	*Graph Theory*
G. N. Watson	*A Treatise on the Theory of Bessel Functions, 2nd Edition*
A. N. Whitehead, B. Russell	*Principia Mathematica to *56, 2nd Edition*
E. T. Whittaker	*A Treatise on the Analytical Dynamics of Particles and Rigid Bodies*
E. T. Whittaker, G. N. Watson	*A Course of Modern Analysis, 4th Edition*
A. Zygmund	*Trigonometric Series, 3rd Edition*

Thermodynamic Formalism

The Mathematical Structures of Equilibrium Statistical Mechanics

Second Edition

DAVID RUELLE

Institut des Hautes Etudes Scientifiques

CAMBRIDGE UNIVERSITY PRESS
Cambridge, New York, Melbourne, Madrid, Cape Town,
Singapore, São Paulo, Delhi, Mexico City

Cambridge University Press
The Edinburgh Building, Cambridge CB2 8RU, UK

Published in the United States of America by Cambridge University Press, New York

www.cambridge.org
Information on this title: www.cambridge.org/9780521546492

First published in *Encyclopedia of Mathematics and Its Applications* 1978
Second edition published 2004

A catalogue record for this publication is available from the British Library

ISBN 978-0-521-54649-2 Paperback

We haven’t seen everything yet,
but when we do
it won’t be for the first time
or the last, either.
You know us.

J. VINOGRAD

Contents

Foreword to the first edition page xv
Preface to the first edition xvii
Preface to the second edition xix

Introduction 1
0.1 Generalities 1
0.2 Description of the thermodynamic formalism 3
0.3 Summary of contents 9

1 Theory of Gibbs states 11
1.1 Configuration space 11
1.2 Interactions 12
1.3 Gibbs ensembles and thermodynamic limit 13
1.4 Proposition 14
1.5 Gibbs states 14
1.6 Thermodynamic limit of Gibbs ensembles 15
1.7 Boundary terms 16
1.8 Theorem 18
1.9 Theorem 18
1.10 Algebra at infinity 19
1.11 Theorem (characterization of pure Gibbs states) 20
1.12 The operators $\mathfrak{M}_\Lambda$ 20
1.13 Theorem (characterization of unique Gibbs states) 21
1.14 Remark 22
Notes 23
Exercises 23

2	**Gibbs states: complements**	24
2.1	Morphisms of lattice systems	24
2.2	Example	25
2.3	The interaction $F^*\Phi$	25
2.4	Lemma	26
2.5	Proposition	26
2.6	Remarks	27
2.7	Systems of conditional probabilities	28
2.8	Properties of Gibbs states	29
2.9	Remark	30
	Notes	30
	Exercises	31
3	**Translation invariance. Theory of equilibrium states**	33
3.1	Translation invariance	33
3.2	The function A_Φ	34
3.3	Partition functions	35
3.4	Theorem	36
3.5	Invariant states	39
3.6	Proposition	39
3.7	Theorem	40
3.8	Entropy	42
3.9	Infinite limit in the sense of van Hove	43
3.10	Theorem	43
3.11	Lemma	45
3.12	Theorem	45
3.13	Corollary	47
3.14	Corollary	48
3.15	Physical interpretation	48
3.16	Theorem	49
3.17	Corollary	49
3.18	Approximation of invariant states by equilibrium states	50
3.19	Lemma	50
3.20	Theorem	52
3.21	Coexistence of phases	53
	Notes	54
	Exercises	54

4 Connection between Gibbs states and Equilibrium states 57
4.1 Generalities 57
4.2 Theorem 58
4.3 Physical interpretation 59
4.4 Proposition 59
4.5 Remark 60
4.6 Strict convexity of the pressure 61
4.7 Proposition 61
4.8 $\mathbf{Z}^\nu$-lattice systems and $\mathbf{Z}^\nu$-morphisms 62
4.9 Proposition 62
4.10 Corollary 63
4.11 Remark 63
4.12 Proposition 64
4.13 Restriction of $\mathbf{Z}^\nu$ to a subgroup G 64
4.14 Proposition 65
4.15 Undecidability and non-periodicity 65
Notes 66
Exercises 66

5 One-dimensional systems 69
5.1 Lemma 70
5.2 Theorem 70
5.3 Theorem 71
5.4 Lemma 72
5.5 Proof of theorems 5.2 and 5.3 73
5.6 Corollaries to theorems 5.2 and 5.3 75
5.7 Theorem 76
5.8 Mixing $\mathbf{Z}$-lattice systems 78
5.9 Lemma 78
5.10 Theorem 79
5.11 The transfer matrix and the operator $\mathfrak{L}$ 80
5.12 The function $\psi_>$ 81
5.13 Proposition 81
5.14 The operator $\mathbb{S}$ 82
5.15 Lemma 82
5.16 Proposition 82
5.17 Remark 83
5.18 Exponentially decreasing interactions 83

5.19 The space $\mathcal{F}^\theta$ and related spaces 84
5.20 Proposition 85
5.21 Theorem 85
5.22 Remarks 86
5.23 Lemma 86
5.24 Proposition 87
5.25 Remark 88
5.26 Theorem 88
5.27 Corollary 89
5.28 Zeta functions 89
5.29 Theorem 90
5.30 Remark 93
Notes 93
Exercises 94

6 Extension of the thermodynamic formalism 101
6.1 Generalities 101
6.2 Expansiveness 101
6.3 Covers 102
6.4 Entropy 103
6.5 Proposition 103
6.6 Pressure 104
6.7 Other definitions of the pressure 105
6.8 Properties of the pressure 106
6.9 The action τ^a 107
6.10 Lemma 107
6.11 Lemma 107
6.12 Theorem (variational principle) 108
6.13 Equilibrium states 110
6.14 Theorem 111
6.15 Remark 111
6.16 Commuting continuous maps 112
6.17 Extension to a $\mathbf{Z}^\nu$-action 112
6.18 Results for $\mathbf{Z}^\nu_\geqslant$-actions 113
6.19 Remark 115
6.20 Topological entropy 115
6.21 Relative pressure 115
6.22 Theorem 116

6.23 Corollary 117
Notes 117
Exercises 118

7 Statistical mechanics on Smale spaces 121
7.1 Smale spaces 121
7.2 Example 123
7.3 Properties of Smale spaces 123
7.4 Smale's "spectral decomposition" 124
7.5 Markov partitions and symbolic dynamics 124
7.6 Theorem 125
7.7 Hölder continuous functions 126
7.8 Pressure and equilibrium states 126
7.9 Theorem 128
7.10 Corollary 128
7.11 Remark 128
7.12 Corollary 129
7.13 Corollary 129
7.14 Equilibrium states for *A* not Hölder continuous 130
7.15 Conjugate points and conjugating homeomorphisms 131
7.16 Proposition 132
7.17 Theorem 132
7.18 Gibbs states 133
7.19 Periodic points 133
7.20 Theorem 134
7.21 Study of periodic points by symbolic dynamics 135
7.22 Proposition 135
7.23 Zeta functions 135
7.24 Theorem 137
7.25 Corollary 137
7.26 Expanding maps 138
7.27 Remarks 139
7.28 Results for expanding maps 140
7.29 Markov partitions 140
7.30 Theorem 141
7.31 Applications 141
Notes 143
Exercises 144

Appendix A.1 Miscellaneous definitions and results 146
A.1.1 Order 146
A.1.2 Residual sets 146
A.1.3 Upper semi-continuity 147
A.1.4 Subadditivity 147

Appendix A.2 Topological dynamics 148

Appendix A.3 Convexity 150
A.3.1 Generalities 150
A.3.2 Hahn–Banach theorem 150
A.3.3 Separation theorems 151
A.3.4 Convex compact sets 151
A.3.5 Extremal points 151
A.3.6 Tangent functionals to convex functions 152
A.3.7 Multiplicity of tangent functionals 152

Appendix A.4 Measures and abstract dynamical systems 153
A.4.1 Measures on compact sets 153
A.4.2 Abstract measure theory 154
A.4.3 Abstract dynamical systems 154
A.4.4 Bernoulli shifts 155
A.4.5 Partitions 155
A.4.6 Isomorphism theorems 156

Appendix A.5 Integral representations on convex compact sets 157
A.5.1 Resultant of a measure 157
A.5.2 Maximal measures 158
A.5.3 Uniqueness problem 158
A.5.4 Maximal measures and extremal points 158
A.5.5 Simplexes of measures 159
A.5.6 $\mathbf{Z}^\nu$-invariant measures 159

Appendix B Open problems 160
B.1 Systems of conditional probabilities (Chapter 2) 160
B.2 Theory of phase transitions (Chapter 3) 160
B.3 Abstract measure-theory viewpoint (Chapter 4) 160
B.4 A theorem of Dobrushin (Chapter 5) 160
B.5 Definition of the pressure (Chapter 6) 161

B.6 Shub's entropy conjecture (Chapter 6) 161
B.7 The condition (SS3) (Chapter 7) 161
B.8 Gibbs states on Smale spaces (Chapter 7) 161
B.9 Cohomological interpretation (Chapter 7) 161
B.10 Smale flows (Chapter 7 and Appendix C) 161

Appendix C Flows 162
C.1 Thermodynamic formalism on a metrizable compact set 162
C.2 Special flows 163
C.3 Special flow over a Smale space 163
C.4 Problems 164

Appendix D Update of open problems 165

References 167
Index 172

Foreword to the first edition

Thermodynamics is still, as it always was, at the center of physics, the standard-bearer of successful science. As happens with many a theory, rich in applications, its foundations have been murky from the start and have provided a traditional challenge on which physicists and mathematicians alike have tested their latest skills.

Ruelle's book is perhaps the first entirely rigorous account of the foundations of thermodynamics. It makes heavier demands on the reader's mathematical background than any volume published so far. It is hoped that ancillary volumes in time will be published which will ease the ascent onto this beautiful and deep theory; at present, much of the background material can be gleaned from standard texts in mathematical analysis. In any case, the timeliness of the content shall be ample reward for the austerity of the text.

GIOVANNI GALLAVOTTI
General Editor, Section on Statistical Mechanics
and
GIAN-CARLO ROTA
Editor, Encyclopedia of Mathematics and its Applications.

Preface to the first edition

The present monograph is based on lectures given in the mathematics departments of Berkeley (1973) and of Orsay (1974–1975). My aim has been to describe the mathematical structures underlying the thermodynamic formalism of equilibrium statistical mechanics, in the simplest case of classical lattice spin systems.

The thermodynamic formalism has its origins in physics, but it has now invaded topological dynamics and differentiable dynamical systems, with applications to questions as diverse as the study of invariant measures for an Anosov diffeomorphism (Sinai [3]), or the meromorphy of Selberg's zeta function (Ruelle [7]). The present text is an introduction to such questions, as well as to more traditional problems of statistical mechanics, like that of phase transitions. I have developed the general theory – which has considerable unity – in some detail. I have, however, left aside particular techniques (like that of correlation inequalities) which are important in discussing examples of phase transitions, but should be the object of a special study.

Statistical mechanics extends to systems vastly more general than the classical lattice spin systems discussed here (in particular to quantum systems). One can therefore predict that the theory discussed in this monograph will extend to vastly more general mathematical setups (in particular to non-commutative situations). I hope that the present text may contribute some inspiration to the construction of the more general theories, as well as clarifying the conceptual structure of the existing formalism.

Preface to the second edition

Twenty-five years have elapsed since the first printing of *Thermodynamic Formalism*, and in the meantime a number of significant developments have taken place in the area indicated by the subtitle *The Mathematical Structures of Equilibrium Statistical Mechanics*. Fortunately, our monograph was concerned with basics, which have remained relatively unchanged, so that *Thermodynamic Formalism* remains frequently quoted. In the present re-issue, some misprints have been corrected, and an update on the open problems of Appendix B has been added. We shall now outline briefly some new developments and indicate unsystematically some source material for these developments. The mathematical aspects of the statistical mechanics of lattice systems, including phase transitions, are covered in the monographs of Sinai [a], and Simon [b]. It may be mentioned that research in this important domain has become less active than it was in the 1960s, '70s, and '80s (but a really good idea might reverse this evolution again). The relation between Gibbs and equilibrium states has been extended to more general topological situations (see Haydn and Ruelle [c]). For a connection of Gibbs states with non-commutative algebras and K-theory, see for instance [d] and the references given there, in particular to the work of Putnam. Particularly fruitful developments have taken place which use the concepts of transfer operators and dynamical zeta functions. In the present monograph these concepts are introduced (in Chapters 5 and 7) in a situation corresponding to uniformly hyperbolic smooth dynamics (Anosov and Axiom A systems, here presented in the topological setting of Smale spaces). The hyperbolic orientation has led to very interesting results concerning the distribution of periods of periodic orbits for hyperbolic flows (in particular the lengths of geodesics on a manifold of negative curvature). These results have been beautifully presented in the monograph of Parry and Pollicott [e]. More recent results of Dolgopyat on exponential decay of correlation for hyperbolic flows [f,g] may be mentioned at this point. It was realized by Baladi and Keller that the ideas of transfer operators and dynamical zeta functions work well also for piecewise monotone maps of the interval (which are not uniformly hyperbolic dynamical systems). This new development can in particular be related to the kneading theory of Thurston and Milnor. We have thus now a much more general theory of transfer operators, very usefully presented in the monograph of Baladi [h], which has in particular an extensive bibliography of the subject. For a general presentation of dynamical zeta functions see also Ruelle [i]. As we have seen, the ideas of statistical mechanics, of a rather algebraic nature, have found geometric applications in smooth dynamics, and particularly the study of hyperbolic systems. Extensions to nonuniformly hyperbolic dynamical systems are currently an active domain of research, with SRB states playing an important role for Sinai, Ruelle, Bowen, Strelcyn, Ledrappier, Young, Viana, . . .). This, however, is another story.

References

[a] Y. G. Sinai. *Phase Transitions: Rigorous Results*. Pergamon Press, Oxford, 1982.

[b] B. Simon. *The Statistical Mechanics of Lattice Gases I*. Princeton University Press, Princeton, 1993.

[c] N. T. A. Haydn and D. Ruelle. "Equivalence of Gibbs and equilibrium states for homeomorphisms satisfying expansiveness and specification," *Commun. Math. Phys.* **148**, 155–167 (1992).

[d] A. Kumjan and D. Pask. "Actions of $\mathbf{Z}^k$ associated to higher rank graphs," *Ergod. Th. and Dynam. Syst.* **23**, 1153–1172 (2003).

[e] W. Parry and M. Pollicott. *Zeta Functions and the Periodic Orbit Structure of Hyperbolic Dynamics*. Astérisque 187–188, Soc. Math. de France, Paris, 1990.

[f] D. Dolgopyat. "On decay of correlations in Anosov flows," *Ann. of Math.* **147**, 357–390 (1998).

[g] D. Dolgopyat. "Prevalence of rapid mixing for hyperbolic flows," *Ergod. Th. and Dynam. Syst.* **18**, 1097–1114 (1998).

[h] V. Baladi. *Positive Transfer Operators and Decay of Correlations*. World Scientific, Singapore, 2000.

[i] D. Ruelle. "Dynamical zeta functions and transfer operators," *Notices Amer. Math. Soc.* **49**, 887–895 (2002).

Introduction

0.1 Generalities

The formalism of equilibrium statistical mechanics – which we shall call *thermodynamic formalism* – has been developed since G. W. Gibbs to describe the properties of certain physical systems. These are systems consisting of a large number of subunits (typically 10^{27}) like the molecules of one liter of air or water. While the physical justification of the thermodynamic formalism remains quite insufficient, this formalism has proved remarkably successful at explaining facts.

In recent years it has become clear that, underlying the thermodynamic formalism, there are mathematical structures of great interest: the formalism hints at the good theorems, and to some extent at their proofs. Outside of statistical mechanics proper, the thermodynamic formalism and its mathematical methods have now been used extensively in *constructive quantum field theory** and in the study of certain *differentiable dynamical systems* (notably Anosov diffeomorphisms and flows). In both cases the relation is at an abstract mathematical level, and fairly inobvious at first sight. It is evident that the study of the physical world is a powerful source of inspiration for mathematics. That this inspiration can act in such a detailed manner is a more remarkable fact, which the reader will interpret according to his own philosophy.

The main physical problem which equilibrium statistical mechanics tries to clarify is that of phase transitions. When the temperature of water is lowered, why do its properties change first smoothly, then suddenly as the freezing point is reached? While we have some general ideas about this, and many special results,

* See for instance Velo and Wightman [1].

a conceptual understanding is still missing.[†] The mathematical investigation of the thermodynamic formalism is in fact not completed; the theory is a young one, with emphasis still more on imagination than on technical difficulties. This situation is reminiscent of preclassic art forms, where inspiration has not been castrated by the necessity to conform to standard technical patterns. We hope that some of this juvenile freshness of the subject will remain in the present monograph!

The physical systems to which the thermodynamic formalism applies are idealized to be actually infinite, i.e. to fill $\mathbf{R}^\nu$ (where $\nu = 3$ in the usual world). This idealization is necessary because only infinite systems exhibit sharp phase transitions. Much of the thermodynamic formalism is concerned with the study of *states* of infinite systems.

For *classical systems* the states are probability measures on an appropriate space of infinite configurations; such states can also be viewed as linear functionals on an abelian algebra (an algebra of continuous functions in the case of Radon measures). For *quantum systems* the states are "expectation value" linear functionals on non-abelian algebras. Due to their greater simplicity, classical systems have been studied more than quantum systems. In fact attention has concentrated on the simplest systems, the *classical lattice systems* where $\mathbf{R}^\nu$ is replaced by $\mathbf{Z}^\nu$ (a ν-dimensional crystal lattice). For such systems the configuration space is a subset Ω of $\prod_{x \in \mathbf{Z}^\nu} \Omega_x$ (where Ω_x is for instance the set of possible "spin values" or "occupation numbers" at the lattice site x). We shall assume that Ω_x is finite. Due to the group invariance (under $\mathbf{Z}^\nu$ or $\mathbf{R}^\nu$) the study of states of infinite systems is closely related to ergodic theory. There are however other parts of the thermodynamic formalism concerned with quite different questions (like analyticity problems).

The present monograph addresses itself to mathematicians. Its aim is to give an account of part of the thermodynamic formalism, and of the corresponding structures and methods. We have restricted ourselves to classical lattice systems. The thermodynamic formalism extends to many other classes of systems, but the theory as it exists now for those systems is less complete, more singular, and filled with technical difficulties. The formalism which we shall describe would not apply directly to the problems of constructive quantum field theory, but it is appropriate to the discussion of Anosov diffeomorphisms and related dynamical systems.

[†] At a more phenomenological level, a good deal is known about phase transitions and much attention has been devoted to critical points and "critical phenomena"; the latter remain however for the moment inaccessible to rigorous investigations.

The mathematics underlying the thermodynamic formalism consists of general methods and special techniques. We have restricted ourselves in this monograph to the general methods; we hope that a complement on special techniques will be published later. As a rough rule, we have decided that a result was not "general" if it required that the configuration space of the system factorize completely in the form $\Omega = \prod \Omega_x$, where Ω_x is the finite set of "spin values" at the lattice site x. The body of general methods thus defined has considerable unity. As for the special techniques, let us mention the correlation inequalities, the method of integral equations, the Lee-Yang circle theorem, and the Peierls argument. These techniques look somewhat specialized from the general point of view taken in this monograph, but are often extremely elegant. They provide, in special situations, a variety of detailed results of great interest for physics.

0.2 Description of the thermodynamic formalism

The contents of this section are not logically required for later chapters. We describe here, for purposes of motivation and orientation, some of the ideas and results of the thermodynamic formalism.* The reader may go over this material rapidly, or skip it entirely.

I. Finite systems

Let Ω be a non-empty finite set. Given a probability measure σ on Ω we define its *entropy*

$$S(\sigma) = -\sum_{\xi \in \Omega} \sigma\{\xi\} \log \sigma\{\xi\},$$

where it is understood that $t \log t = 0$ if $t = 0$. Given a function $U : \Omega \to \mathbf{R}$, we define a real number Z called the *partition function* and a probability measure ρ on Ω called the *Gibbs ensemble* by

$$Z = \sum_{\xi \in \Omega} \exp[-U(\xi)],$$

$$\rho\{\xi\} = Z^{-1} \exp[-U(\xi)]. \tag{0.1}$$

Proposition (Variational principle). *The maximum of the expression*†

$$S(\sigma) - \sigma(U)$$

* We follow in part the Séminaire Bourbaki, exposé 480.
† We write $\sigma(U) = \sum_\xi \sigma\{\xi\}U(\xi)$ or more generally $\sigma(U) = \int U(\xi)\sigma(d\xi)$.

over all probability measures σ *on* Ω *is* $\log Z$, *and is reached precisely for* $\sigma = \rho$.

For physical applications, Ω is interpreted as the space of configurations of a finite system. One writes $U = \beta E$, where $E(\xi)$ is the energy of the configuration ξ, and $\beta = 1/kT$, where T is the absolute temperature and k is a factor known as Boltzmann's constant. The problem of why the Gibbs ensemble describes thermal equilibrium (at least for "large systems") when the above physical identifications have been made is deep and incompletely clarified. Note that the energy E may depend on physical parameters called "magnetic field," "chemical potential," etc. Note also that the traditional definition of the energy produces a minus sign in $\exp[-\beta E]$, which is in practice a nuisance. From now on we absorb β in the definition of U, and call U the *energy*. We shall retain from the above discussion only the hint that the Gibbs ensemble is an interesting object to consider in the limit of a "large system."

The thermodynamic formalism studies measures analogous to the Gibbs ensemble ρ in a certain limit where Ω becomes infinite, but some extra structure is present. Imitating the variational principle of the above Proposition, one defines *equilibrium states* (see II below). Imitating the definition (0.1), one defines *Gibbs states* (see III below).

II. Thermodynamic formalism on a metrizable compact set

Let Ω be a non-empty metrizable compact set, and $x \to \tau^x$ a homomorphism of the additive group $\mathbf{Z}^\nu (\nu \geqslant 1)$ into the group of homeomorphisms of Ω. We say that τ is *expansive* if, for some allowed metric d, there exists $\delta > 0$ such that

$$(d(\tau^x \xi, \tau^x \eta) \leqslant \delta \quad \text{for all } x) \Rightarrow (\xi = \eta).$$

Definition of the pressure. If $\mathfrak{A} = (\mathfrak{A}_i)$, $\mathfrak{B} = (\mathfrak{B}_j)$ are covers of Ω, the cover $\mathfrak{A} \vee \mathfrak{B}$ consists of the sets $\mathfrak{A}_i \cap \mathfrak{B}_j$. This notation extends to any finite family of covers. We write

$$\tau^{-x}\mathfrak{A} = (\tau^{-x}\mathfrak{A}_i),$$
$$\mathfrak{A}^\Lambda = \bigvee_{x \in \Lambda} \tau^{-x}\mathfrak{A} \qquad \text{if} \quad \Lambda \subset \mathbf{Z}^\nu,$$
$$\operatorname{diam} \mathfrak{A} = \sup_i \operatorname{diam} \mathfrak{A}_i,$$

where $\operatorname{diam} \mathfrak{A}_i$ is the diameter of $\mathfrak{A}_i$ for an allowed metric d on Ω.

The definition of the pressure which we shall now give will not look simple and natural to someone unfamiliar with the subject. This should not alarm the reader: the definition will give us quick access to a general statement of theorems of statistical mechanics. It will otherwise recur only in Chapter 6, with more preparation.

We denote by $\mathcal{C} = \mathcal{C}(\Omega)$ the space of continuous real functions on Ω. Let $A \in \mathcal{C}$, $\mathfrak{A}$ be a finite open cover of Ω, and Λ be a finite subset of $\mathbf{Z}^\nu$; define

$$Z_\Lambda(A, \mathfrak{A}) = \min \left\{ \sum_j \exp \left[\sup_{\xi \in \mathfrak{B}_j} \sum_{x \in \Lambda} A(\tau^x \xi) \right] \right.$$
$$\left. : (\mathfrak{B}_j) \text{ is a subcover of } \mathfrak{A}^\Lambda \right\}.$$

If $a^1, \ldots, a^\nu$ are integers > 0, let $a = (a^1, \ldots, a^\nu)$ and

$$\Lambda(a) = \{(x^1, \ldots, x^\nu) \in \mathbf{Z}^\nu : 0 \leqslant x^i < a^i \text{ for } i = 1, \ldots, \nu\}.$$

The function $a \to \log Z_{\Lambda(a)}(A, \mathfrak{A})$ is subadditive, and one can write (with $|\Lambda(a)| = \text{card}\, \Lambda(a) = \prod_i a^i$)

$$\begin{aligned} P(A, \mathfrak{A}) &= \lim_{a^1, \ldots, a^\nu \to \infty} \frac{1}{|\Lambda(a)|} \log Z_{\Lambda(a)}(A, \mathfrak{A}) \\ &= \inf_a \frac{1}{|\Lambda(a)|} \log Z_{\Lambda(a)}(A, \mathfrak{A}), \end{aligned}$$

and

$$P(A) = \lim_{\text{diam}\, \mathfrak{A} \to 0} P(A, \mathfrak{A}).$$

The function $P : \mathcal{C} \to \mathbf{R} \cup \{+\infty\}$ is the (topological) *pressure*. $P(A)$ is finite for all A if and only if $P(0)$ is finite; in that case P is convex and continuous (for the topology of uniform convergence in $\mathcal{C}$). $P(0)$ is the *topological entropy*; it gives a measure of the rate of mixing of the action τ.

Entropy of an invariant measure. If σ is a probability measure on Ω, and $\mathfrak{A} = (\mathfrak{A}_i)$ a finite Borel partition of Ω, we write

$$H(\sigma, \mathfrak{A}) = -\sum_i \sigma(\mathfrak{A}_i) \log \sigma(\mathfrak{A}_i).$$

The real measures on Ω constitute the dual $\mathcal{C}^*$ of $\mathcal{C}$. The topology of weak dual of $\mathcal{C}$ on $\mathcal{C}^*$ is called the *vague* topology. Let $I \subset \mathcal{C}^*$ be the set of probability measures σ invariant under τ, i.e. such that $\sigma(A) = \sigma(A \circ \tau^x)$; I is convex and compact for the vague topology. If $\mathfrak{A}$ is a finite Borel partition and $\sigma \in I$, we write

$$h(\sigma, \mathfrak{A}) = \lim_{a^1,\dots,a^\nu \to \infty} \frac{1}{|\Lambda(a)|} H(\sigma, \mathfrak{A}^{\Lambda(a)})$$
$$= \inf_a \frac{1}{|\Lambda(a)|} H(\sigma, \mathfrak{A}^{\Lambda(a)});$$
$$h(\sigma) = \lim_{\operatorname{diam} \mathfrak{A} \to 0} h(\sigma, \mathfrak{A}).$$

The function $h : I \to \mathbf{R} \cup \{+\infty\}$ is affine $\geqslant 0$; it is called the (mean) *entropy*. If τ is expansive, h is finite and upper semi-continuous on I (with the vague topology).

Theorem 1 (Variational principle).

$$P(A) = \sup_{\sigma \in I} [h(\sigma) + \sigma(A)]$$

for all $A \in C$.

This corresponds to the variational principle for finite systems if $-A$ is interpreted as the contribution to the energy of one lattice site.

Let us assume that P is finite. The set I_A of *equilibrium states* for $A \in \mathcal{C}$ is defined by

$$I_A = \{\sigma \in I : h(\sigma) + \sigma(A) = P(A)\}.$$

I_A may be empty.

Theorem 2 *Assume that h is finite and upper semi-continuous on I (with the vague topology).*

(a) $I_A = \{\sigma \in \mathcal{C}^* : P(A + B) \geqslant P(A) + \sigma(B)$ for all $B \in \mathcal{C}\}$. *This set is not empty; it is convex, compact; it is a Choquet simplex and a face of I.*
(b) *The set* $D = \{A \in \mathcal{C} : \text{card } I_A = 1\}$ *is residual in* $\mathcal{C}$.
(c) *For every* $\sigma \in I$,

$$h(\sigma) = \inf_{A \in \mathcal{C}} [P(A) - \sigma(A)].$$

The fact that I_A is a metrizable simplex implies that each $\sigma \in I_A$ has a unique integral representation as the barycenter of a measure carried by the extremal points of I_A. It is known that I is also a simplex. The fact that I_A is a face of I implies that the extremal points of I_A are also extremal points of I (i.e. τ-ergodic measures on Ω).

III. Statistical mechanics on a lattice

The above theorems extend results known for certain systems of statistical mechanics (classical lattice systems). For instance, if F is a non-empty finite set (with the discrete topology), we can take $\Omega = F^{\mathbf{Z}^\nu}$ with the product topology, and τ^x defined in the obvious manner. More generally we shall take for Ω a closed τ-invariant non-empty subset of $F^{\mathbf{Z}^\nu}$. For the physical interpretation, note that Ω is the space of infinite configurations of a system of spins on a crystal lattice $\mathbf{Z}^\nu$. Up to sign and factors of β, P can be interpreted as the "free energy" or the "pressure," depending on the physical interpretation of F as the set of "spin values" or of "occupation numbers" at a lattice site. For simplicity we have retained the word "pressure."

If $x = (x^i) \in \mathbf{Z}^\nu$, we write $|x| = \max |x^i|$. Let $0 < \lambda < 1$; if $\xi, \eta \in \Omega$, with $\xi = (\xi_x)_{x \in \mathbf{Z}^\nu}$, $\eta = (\eta_x)_{x \in \mathbf{Z}^\nu}$, we define

$$d(\xi, \eta) = \lambda^k \qquad \text{with} \quad k = \inf\{|x| : \xi_x \neq \eta_x\}.$$

d is a distance compatible with the topology of Ω. One checks with this definition that τ is expansive; hence Theorem 2 applies.

We shall henceforth assume that there exists a finite set $\Delta \subset \mathbf{Z}^\nu$ and $G \subset F^\Delta$ such that

$$\Omega = \{\xi \in F^{\mathbf{Z}^\nu} : \tau^x \xi | \Delta \in G \qquad \text{for all} \quad x\}.$$

If $\Lambda \subset \mathbf{Z}^\nu$ we denote by pr_Λ, pr'_Λ the projections of $F^{\mathbf{Z}^\nu}$ on F^Λ and $F^{\mathbf{Z}^\nu \backslash \Lambda}$ respectively.

Let $0 < \alpha \leqslant 1$, and denote by $\mathcal{C}^\alpha$ the Banach space of real Hölder continuous functions of exponent α on Ω (with respect to the metric d). Let $\xi = (\xi_x) \in \Omega$, $\eta = (\eta_x) \in \Omega$. If $\xi_x = \eta_x$ except for finitely many x, and $A \in \mathcal{C}^\alpha$, we can write

$$g_A(\xi, \eta) = \prod_{x \in \mathbf{Z}^\nu} \exp[A(\tau^x \xi) - A(\tau^x \eta)]$$

because $|A(\tau^x \xi) - A(\tau^x \eta)| \to 0$ exponentially fast when $|x| \to \infty$. For finite Λ, a continuous function $f_\Lambda : \mathrm{pr}_\Lambda \Omega \times \mathrm{pr}'_\Lambda \Omega \to \mathbf{R}$ is then defined by

$$f_\Lambda(\xi) = \begin{cases} \left[\sum_{\eta \in \Omega : \mathrm{pr}'_\Lambda \eta = \mathrm{pr}'_\Lambda \xi} g_A(\eta, \xi) \right]^{-1} & \text{if} \quad \xi \in \Omega, \\ 0 & \text{if} \quad \xi \notin \Omega. \end{cases}$$

Definition. Let $A \in \mathcal{C}^\alpha$; we say that a probability measure σ on Ω is a *Gibbs state* if the following holds.

For every finite $\Lambda \subset \mathbf{Z}^\nu$, let ϵ_Λ be the measure on $\mathrm{pr}_\Lambda \Omega$ which gives to each point of this set the mass 1. Then

$$\sigma = f_\Lambda \cdot (\epsilon_\Lambda \otimes \mathrm{pr}'_\Lambda \sigma);$$

(we have again denoted by σ the image of this measure by the canonical map $\Omega \mapsto \mathrm{pr}_\Lambda \Omega \times \mathrm{pr}'_\Lambda \Omega$).

Another formulation of the definition is this: σ is a Gibbs state if, for every finite Λ, the conditional probability that $\xi|\Lambda$ is realized in Λ, knowing that $\xi|(\mathbf{Z}^\nu \backslash \Lambda)$ realized in $\mathbf{Z}^\nu \backslash \Lambda$, is $f_\Lambda(\xi)$.

Theorem 3 *Let $A \in \mathcal{C}^\alpha$.*

(a) *Every equilibrium state is a τ-invariant Gibbs state.*
(b) *If $\Omega = F^{\mathbf{Z}^\nu}$, every τ-invariant Gibbs state is an equilibrium state.*

In view of (a), the Gibbs states are the probability measures which have the same conditional probabilities f_Λ as the equilibrium states. Part (b) of the theorem holds under conditions much more general than $\Omega = F^{\mathbf{Z}^\nu}$. The assumption $A \in \mathcal{C}^\alpha$ can also be considerably weakened. For simplicity we have in this section made an unusual description of statistical mechanics, using (Hölder continuous) functions on Ω, rather than the "interactions" which are much more appropriate to a detailed study.

Theorem 4 *The set of Gibbs states for $A \in \mathcal{C}^\alpha$ is a Choquet simplex.*

Every Gibbs state has thus a unique integral decomposition in terms of extremal (or "pure") Gibbs states.

Physical interpretation. The extremal equilibrium states are τ-ergodic measures. They are interpreted as *pure thermodynamic phases.* Since the equilibrium states correspond to tangents to the graph of P (Theorem 2(a)), the discontinuities of the derivative of P correspond to *phase transitions.* One would thus like to know if P is piecewise analytic (in a suitable sense) on $\mathcal{C}^\alpha$. An extremal equilibrium state σ may have a non-trivial decomposition into extremal Gibbs states (those will not be τ-invariant, because of Theorem 3(b)). This is an example of *symmetry breaking* (the broken symmetry is the invariance under τ).

The main problem of equilibrium statistical mechanics is to understand the nature of phases and phase transitions. Because of this, the main object of the thermodynamic formalism is to study the differentiability and analyticity properties of the function P, and the structure of the equilibrium states and Gibbs states. As already mentioned, detailed results are known only in special cases, but we shall restrict ourselves in the present monograph to the general theory, as it is known at the time of writing.

For "one-dimensional systems," i.e. for $\nu = 1$, there are fairly complete results, which can be summarized by saying that there are no phase transitions. Let us assume that

$$\Omega = \{\xi = (\xi_x)_{x\in\mathbf{Z}} \in F^{\mathbf{Z}} : t_{\xi_x \xi_{x+1}} = 1 \qquad \text{for all} \quad x\},$$

where $t = (t_{uv})$ is a matrix with elements 0 or 1. We assume also that there exists an integer $N > 0$ such that all the matrix elements of t^N are > 0.

Theorem 5 *If the above conditions are satisfied,* $P : \mathcal{C}^\alpha \to \mathbf{R}$ *is real analytic. Furthermore for every* $A \in \mathcal{C}^\alpha$ *there is only one Gibbs state (which is also the only equilibrium state).*

All these properties are false for $\nu > 1$.

0.3 Summary of contents

Chapters 1 to 5 of this monograph are devoted to the general theory of equilibrium statistical mechanics of classical lattice systems; complete proofs are generally given. Chapters 6 and 7 extend the thermodynamic formalism outside of the traditional domain of statistical mechanics: here the proofs are largely omitted or only sketched.* We give now some more details.

Chapters 1 and 2 give the theory of Gibbs states, without assuming invariance under lattice translations (the lattice $\mathbf{Z}^\nu$ is thus replaced by a general infinite countable set L). Chapter 3 assumes translation invariance and develops the theory of equilibrium states and of the pressure for classical lattice systems; general results on phase transitions are also obtained. Chapter 4 is central, and establishes the connexion between Gibbs states and equilibrium states. Chapter 5 deals with one-dimensional systems and prepares Chapter 7. Chapter 6 extends the theory of equilibrium states to the situation where the configuration space Ω is replaced by a general compact metrizable space on which $\mathbf{Z}^\nu$ acts by homeomorphisms. Chapter 7 extends the theory of Gibbs states (and related topics) to a certain class of compact metric spaces, which we call *Smale spaces*, on which $\mathbf{Z}$ acts by homeomorphisms. Smale spaces include Axiom A basic sets and in particular manifolds with an Anosov diffeomorphism.

Some extra material is given in the form of exercises at the end of the chapters.

Bibliographic references are given either in the text or in notes at the end of the chapters. For purposes of orientation, it may be good to read these notes

* Of course references to the literature are indicated as needed.

before the corresponding chapter. The reader is particularly advised to consult the following original papers: Ruelle [1], Dobrushin [2], [3], Lanford and Ruelle [1], Israel [1], and Sinai [4].

Some background material has been collected in Appendices A.1 to A.5. These appendices recall some well-known facts to establish terminology, and also provide access to less standard results. In general the reader is assumed to be familiar with basic facts of functional analysis, but no knowledge of physics is presupposed.

A few open problems are collected in Appendix B. Appendix C contains a brief introduction to flows.

Concerning notation and terminology we note the following points. We shall often write $|X|$ for the cardinality of a finite set X. We shall use in Chapters 5–7 the notation $\mathbf{Z}_>, \mathbf{Z}_\geqslant, \mathbf{Z}_<, \mathbf{Z}_\leqslant$ for the sets of integers which are respectively >0, $\geqslant 0$, <0, $\leqslant 0$. A measure ρ will (unless otherwise indicated) be a *Radon measure* on a compact set Ω. If $f : \Omega \to \Omega'$ is a continuous map, the image of ρ by f (see Appendix A.4) is denoted by $f\rho$ (*not* $f^*\rho$).

We refer the reader to Ruelle [3] for a wider study of equilibrium statistical mechanics, and to the excellent monograph by Bowen [6] for applications to differentiable dynamical systems.* Let us also mention the monograph by Israel [2] and the notes by Lanford [2], Georgii [1], and Preston [1], [2]. Monographs are planned by various authors on aspects of statistical mechanics not covered here, but at this time, much interesting material is not available in book form.

Before proceeding with Chapter 1, the reader is invited to go rapidly through the Appendices A.1–A.5.

* For modern introductions to ergodic theory and topological dynamics, see Walters [2]: Denker, Grillenberger, and Sigmund [1].

1
Theory of Gibbs states

This chapter is devoted to the general theory of Gibbs states. No invariance under translations is assumed.

1.1 Configuration space

The following are supposed to be given:

L : an infinite countable set;
Ω_x: a finite set for each $x \in L$;
$\mathfrak{F}$: a set of finite subsets of L, which is locally finite (i.e. each $x \in L$ is an element of only a finite number of the $\Lambda \in \mathfrak{F}$);

$(\bar{\Omega}_\Lambda)_{\Lambda \in \mathfrak{F}}$: a family where $\bar{\Omega}_\Lambda \subset \prod_{x \in \Lambda} \Omega_x$.

We can then define a *configuration space*

$$\Omega = \left\{ \xi \in \prod_{x \in L} \Omega_x : (\forall \Lambda \in \mathfrak{F})\, \xi | \Lambda \in \bar{\Omega}_\Lambda \right\}. \tag{1.1}$$

We shall always assume that $\Omega \neq \emptyset$.

An important example is when $\mathfrak{F}$ is empty. In this case $\Omega = \prod_{x \in L} \Omega_x$.

It will be convenient to think of L as describing a crystal lattice. At each *site* $x \in L$ the system may be in a finite number of different states $\xi_x \in \Omega_x$. For instance, in the description of an alloy, Ω_x is the list of atom species which may occur at x. For a spin system Ω_x is the list of possible spin orientations of the atom at x. One frequently considers a *lattice gas* with $\Omega_x = \{1, 0\}$ (the site x is occupied by an atom, or empty) and a *spin system* with $\Omega_x = \{1, -1\}$ (spin up and spin down). A *configuration* of our system is given by an element $\xi = (\xi_x)_{x \in L}$ of $\prod_{x \in L} \Omega_x$. We impose a certain number of conditions $(\xi_x)_{x \in \Lambda} \in \bar{\Omega}_\Lambda$. For a lattice gas an example of such conditions would be that two "neighboring" lattice sites cannot be occupied simultaneously (the particles on the lattice have *hard cores*).

For any set $S \subset L$, let

$$\Omega_S = \left\{ \xi \in \prod_{x \in S} \Omega_x : (\forall \Lambda \in \mathcal{F} : \Lambda \subset S) \xi | \Lambda \in \bar{\Omega}_\Lambda \right\}. \tag{1.2}$$

Since $\Omega \neq \emptyset$, no Ω_S is empty. We put on Ω_x the discrete topology and on $\prod_{x \in S} \Omega_x$ the product topology, which is compact. The spaces Ω_S, in particular Ω, are then also compact. We have continuous maps $\alpha_S : \Omega \mapsto \Omega_S$ defined by $\alpha_S \xi = \xi | S$ or $\alpha_S(\xi_x)_{x \in L} = (\xi_x)_{x \in S}$. More generally we can define $\alpha_{TS} : \Omega_S \mapsto \Omega_T$ if $S \supset T$. Let $\mathcal{C} = \mathcal{C}(\Omega)$ be the algebra of real continuous functions on Ω. $\mathcal{C}$ is a Banach space with respect to the uniform norm, and the probability measures on Ω form a convex compact subset E of the weak dual $\mathcal{C}^*$ of $\mathcal{C}$ ($\mathcal{C}^*$ is the space of real measures on Ω; the topology of weak dual is the *vague topology*). E is metrizable. For finite $\Lambda \subset L$, let $\mathcal{C}_\Lambda$ be the algebra of functions $A \circ \alpha_\Lambda$ where $A \in \mathcal{C}(\Omega_\Lambda)$. By the Stone–Weierstrass theorem the union of the $\mathcal{C}_\Lambda$ is dense in $\mathcal{C}$.

We may consider the elements of $\mathcal{C}$ as *observables* of our classical lattice system. The elements of $\mathcal{C}_\Lambda$ are then the physical quantities which can be observed in the *finite region* Λ. A probability measure $\mu \in E$ is a *state*. We interpret it as an expectation value functional on the observables, i.e. a positive linear functional on $\mathcal{C}$ such that $\mu(1) = 1$.

1.2 Interactions

An *interaction* is a real function Φ on

$$\bigcup_{\Lambda \text{ finite} \subset L} \Omega_\Lambda$$

such that $\Phi | \Omega_\emptyset = 0$ and for each $x \in L$

$$|\Phi|_x = \sum_{X \ni x} \frac{1}{|X|} \sup_{\xi \in \Omega_x} |\Phi(\xi)| < +\infty, \tag{1.3}$$

where $|X| = \text{card } X$. Given Φ, an *energy* function $U_\Lambda^\Phi : \Omega_\Lambda \mapsto \mathbf{R}$ is defined for each finite $\Lambda \subset L$ by

$$U_\Lambda^\Phi(\xi) = \sum_{X \subset \Lambda} \Phi(\xi | X). \tag{1.4}$$

We write U_Λ instead of U_Λ^Φ when there is no confusion. We have

$$U_\Lambda(\xi) = \sum_{x\in\Lambda} \sum_{X:x\in X\subset\Lambda} \frac{1}{|X|}\Phi(\xi\,|X). \tag{1.5}$$

Therefore

$$|U_\Lambda| \leqslant \sum_{x\in\Lambda} |\Phi|_x. \tag{1.6}$$

For the rest of this chapter we shall impose instead of (1.3) the stronger requirement

$$\|\Phi\|_x = \sum_{X\ni x} \sup_{\xi\in\Omega_x} |\Phi(\xi)| < +\infty. \tag{1.7}$$

If Λ, M are disjoint subsets of L such that Λ is finite and $\xi \in \Omega_{\Lambda\cup M}$, we can define

$$W_{\Lambda M}(\xi) = \sum_X{}^* \Phi(\xi\,|X), \tag{1.8}$$

where $\sum^*$ extends over the finite sets $X \subset \Lambda \cup M$ such that $X \cap \Lambda \neq \emptyset$ and $X \cup M \neq \emptyset$. We have thus

$$|W_{\Lambda M}| \leqslant \sum_{x\in\Lambda} \|\Phi\|_x. \tag{1.9}$$

If Λ, M are finite, we have also

$$U_{\Lambda\cup M} = U_\Lambda + U_M + W_{\Lambda M}. \tag{1.10}$$

1.3 Gibbs ensembles and thermodynamic limit

The *Gibbs ensemble* for the region Λ (Λ finite $\subset L$) and the interaction Φ is the probability measure $\mu_{(\Lambda)}$ on Ω_Λ defined by

$$\begin{aligned} \mu_{(\Lambda)}\{\xi\} &= Z_\Lambda^{-1} \exp\left[-U_\Lambda^\Phi(\xi)\right], \\ Z_\Lambda &= \sum_{\xi\in\Omega_\Lambda} \exp\left[-U_\Lambda^\Phi(\xi)\right]. \end{aligned} \tag{1.11}$$

The general name of *thermodynamic limit* is given to a limit when Λ "tends to infinity." In particular we write $\Lambda \to L$ if eventually $\Lambda \supset \Delta$ for every finite $\Delta \subset L$ (limit on the increasing net of finite subsets of L ordered by inclusion).

We shall now prove the existence of thermodynamic limits for the Gibbs ensembles by a compactness argument.

1.4 Proposition

Let (M_n) be a sequence of finite subsets of L such that $M_n \to L$, and for each n let $\mu_{(M_n)}$ be a probability measure on Ω_{M_n}. One can choose a subsequence (M_n') such that the following limit exists for all finite $\Lambda \subset L$:

$$\lim_{n\to\infty} \alpha_{\Lambda M_n'} \, \mu_{(M_n')} = \rho_\Lambda. \tag{1.12}$$

Furthermore there exists a unique probability measure ρ on Ω such that

$$\rho_\Lambda = \alpha_\Lambda \rho \tag{1.13}$$

for all Λ.

Notice that $\alpha_{\Lambda M_n}\mu_{(M_n)}$ is defined only for $M_n \supset \Lambda$, i.e. for n sufficiently large. Since the finite subsets Λ of L form a countable family, and the Ω_Λ are finite, the existence of a subsequence (M_n') such that (1.12) holds follows from a diagonal argument. As a consequence of (1.12) we have $\alpha_{\Lambda M}\rho_M = \rho_\Lambda$ if $\Lambda \subset M$. Therefore we can define consistently $\rho(A \circ \alpha_\Lambda) = \rho_\Lambda(A)$ for $A \in \mathcal{C}(\Omega_\Lambda)$ and extend ρ by continuity to $\mathcal{C}$.* The uniqueness of ρ follows from the density of $\cup_\Lambda \mathcal{C}_\Lambda$ in $\mathcal{C}$. We shall say that ρ is a *thermodynamic limit* of the probability measures $\mu_{(\Lambda)}$.

No use has been made of the fact that the probability measures $\mu_{(\Lambda)}$ were Gibbs ensembles. If this is the case, the thermodynamic limit ρ has however special properties, as we shall see in Section 1.6.

1.5 Gibbs states

We say that $\sigma \in E$ is a *Gibbs state* (for the interaction Φ) if, for all finite $\Lambda \subset L$, there exists a probability measure $\sigma_{L\setminus\Lambda}$ on $\Omega_{L\setminus\Lambda}$ such that for all $\xi_\Lambda \in \Omega_\Lambda$

$$(\alpha_\Lambda\sigma)\{\xi_\Lambda\} = \int_{\Omega_{L\setminus\Lambda}} \sigma_{L\setminus\Lambda}(d\eta)\mu_{(\Lambda)\eta}\{\xi_\Lambda\}, \tag{1.14}$$

where

$$\mu_{(\Lambda)_\eta}\{\xi_\Lambda\} = \frac{e^{-U_\Lambda(\xi_\Lambda)-W_{\Lambda,L\setminus\Lambda}(\xi_\Lambda\vee\eta)}}{\sum_{\eta_\Lambda\in\Omega_\Lambda} e^{-U_\Lambda(\eta_\Lambda)-W_{\Lambda,L\setminus\Lambda}(\eta_\Lambda\vee\eta)}}. \tag{1.15}$$

In this formula $\xi_\Lambda \vee \eta$ is the element ζ of Ω such that $\zeta|\Lambda = \xi_\Lambda$, $\zeta|(L\setminus\Lambda) = \eta$, and it is understood that terms with $\xi_\Lambda \vee \eta$ or $\eta_\Lambda \vee \eta$ not defined are omitted.

* Given Λ one can choose M finite such that $\alpha_{\Lambda M}\Omega_M = \alpha_\Lambda\Omega$. Then $|\rho(A \circ \alpha_\Lambda)| = |\rho_\Lambda(A)| = |\rho_M(A \circ \alpha_{\Lambda M})| \leqslant 1\|A \circ \alpha_{\Lambda M}\| = \|A \circ \alpha_\Lambda\|$.

This means that if $\xi_\Lambda \vee \eta$ is not defined we put $\exp[-W_{\Lambda,L\setminus\Lambda}(\xi_\Lambda \vee \eta)] = 0$. We also put the fraction (1.15) equal to zero if its numerator vanishes. Notice that, since $\mathfrak{F}$ has been assumed locally finite, the sets $\{\eta \in \Omega_{L\setminus\Lambda} : \xi_\Lambda \vee \eta \in \Omega\}$ and $\{\eta \in \Omega_{L\setminus\Lambda} : \xi_\Lambda \vee \eta$ not defined$\}$ are both open. On the other hand $W_{\Lambda,L\setminus\Lambda}$ is on $\{\eta \in \Omega_{L\setminus\Lambda} : \xi_\Lambda \vee \eta \in \Omega\}$ a uniformly convergent sum of continuous functions (see (1.7), (1.8)). Therefore *the functions* $\eta \mapsto \exp[-W_{\Lambda,L\setminus\Lambda}(\xi_\Lambda \vee \eta)]$, $\eta \mapsto \mu_{(\Lambda)\eta}\{\xi_\Lambda\}$ *are continuous on* $\Omega_{L\setminus\Lambda}$.

1.6 Thermodynamic limit of Gibbs ensembles

We prove now that if ρ is a thermodynamic limit of Gibbs ensembles $\mu_{(\Lambda)}$ for the interaction Φ, then ρ is a Gibbs state for the interaction Φ.

Using the definition (1.11) we have, for $\Lambda \subset M$,

$$\begin{aligned}(\alpha_{\Lambda M}\mu_{(M)})\{\xi_\Lambda\} &= \sum_{\eta\in\Omega_{M\setminus\Lambda}} \mu_{(M)}\{\xi_\Lambda \vee \eta\} \\ &= \sum_{\eta\in\Omega_{M\setminus\Lambda}} \left[Z_M^{-1} e^{-U_{M\setminus\Lambda}(\eta)}\right] \exp[-U_\Lambda(\xi_\Lambda) - W_{\Lambda,M\setminus\Lambda}(\xi_\Lambda \vee \eta)] \\ &= \sum_{\eta\in\Omega_{M\setminus\Lambda}} [(\alpha_{M\setminus\Lambda,M}\mu_{(M)})\{\eta\}]\mu_{(\Lambda,M)\eta}\{\xi_\Lambda\}, \end{aligned} \tag{1.16}$$

where

$$\mu_{(\Lambda,M)\eta}\{\xi_\Lambda\} = \frac{e^{-U_\Lambda(\xi_\Lambda)-W_{\Lambda,M\setminus\Lambda}(\xi_\Lambda\vee\eta)}}{\sum_{\eta_\Lambda\in\Omega_\Lambda} e^{-U_\Lambda(\eta_\Lambda)-W_{\Lambda,M\setminus\Lambda}(\eta_\Lambda\vee\eta)}}. \tag{1.17}$$

We adopt conventions similar to those of Section 1.5 concerning the meaning of $\xi_\Lambda \vee \eta$. In particular we put the expression (1.17) equal to zero when the numerator vanishes. We now investigate what happens when we replace M in (1.16) by the sequence (M'_n) of Proposition 1.4, and let $n \to \infty$.

(a) We notice that the function on $\Omega_{L\setminus\Lambda}$ defined by

$$\eta \mapsto \mu_{(\Lambda,M'_n)(\eta|M'_n\setminus\Lambda)}\{\xi_\Lambda\}$$

converges uniformly to $\eta \mapsto \mu_{(\Lambda)\eta}\{\xi_\Lambda\}$ because of the convergence of the sum defining $W_{\Lambda,L\setminus\Lambda}(\xi_\Lambda \vee \eta)$.

(b) Using Proposition 1.4 for the sequence $\alpha_{M'_n\setminus\Lambda,M_n}\mu_{(M'_n)}$, we see that it has a thermodynamic limit $\rho_{L\setminus\Lambda}$, which is a probability measure on $\Omega_{L\setminus\Lambda}$.

From (a), (b), and (1.16) we obtain

$$\lim_{n\to\infty} (\alpha_{\Lambda M'_n}\mu_{(M'_n)})\{\xi_\Lambda\} = \int_{\Omega_{L\backslash\Lambda}} \rho_{L\backslash\Lambda}(d\eta)\mu_{(\Lambda)\eta}\{\xi_\Lambda\}.$$

In view of Proposition 1.4 this expresses that ρ is a Gibbs state.

1.7 Boundary terms

Let a probability measure $\mu'_{(\Lambda)}$ on Ω_Λ be defined for each finite $\Lambda \subset L$, or for a sequence (Λ_n) such that $\Lambda_n \to \infty$, or for a suitable net. Suppose that $\mu'_{(\Lambda)}$ is of the form

$$\mu'_{(\Lambda)}\{\xi_\Lambda\} = Z'^{-1}_\Lambda e^{-U_\Lambda(\xi_\Lambda)-B_\Lambda(\xi_\Lambda)},$$
$$Z'_\Lambda = \sum_{\eta_\Lambda\in\Omega_\Lambda} e^{-U_\Lambda(\eta_\Lambda)-B_\Lambda(\eta_\Lambda)}.$$

If the *boundary term* B_Λ has a suitable behavior in the thermodynamic limit, the results of Section 1.6 can be extended to $(\mu'_{(\Lambda)})$.

Instead of (1.16) we have here

$$\begin{aligned}(\alpha_{\Lambda M}\mu'_{(M)})\{\xi_\Lambda\} &= \sum_{\eta\in\Omega_{M\backslash\Lambda}} \mu'_{(M)}\{\xi_\Lambda \vee \eta\} \\ &= \sum_{\eta\in\Omega_{M\backslash\Lambda}} \left[Z'^{-1}_M e^{-U_{M\backslash\Lambda}(\eta)-B'(\eta)}\right] \\ &\quad\times \exp\left[-U_\Lambda(\xi_\Lambda) - W_{\Lambda,M\backslash\Lambda}(\xi_\Lambda\vee\eta) - B''(\xi_\Lambda,\eta)\right]\end{aligned} \tag{1.18}$$

provided $B_M(\xi_\Lambda \vee \eta) = B'(\eta) + B''(\xi_\Lambda, \eta)$. Corresponding to (a) of Section 1.6, suppose that the function on $\Omega_{L\backslash\Lambda}$ defined by

$$\eta \mapsto \exp\left[-U_\Lambda(\xi_\Lambda) - W_{\Lambda,M\backslash\Lambda}(\xi_\Lambda \vee (\alpha_{M\backslash\Lambda,L\backslash\Lambda}\eta)) - B''(\xi_\Lambda, \alpha_{M\backslash\Lambda,L\backslash\Lambda}\eta)\right] \tag{1.19}$$

tends uniformly to $\eta \mapsto \exp[-U_\Lambda(\xi_\Lambda) - W_{\Lambda,L\backslash\Lambda}(\xi_\Lambda \vee \eta)]$; then the argument in Section 1.6 carries over to the present case, and a thermodynamic limit of $(\mu'_{(\Lambda)})$ is a Gibbs state.

We now consider an example. For each finite $\Lambda \subset L$, let $\eta_{L\backslash\Lambda}$ be the restriction to $L\backslash\Lambda$ of some $\eta^*_{(\Lambda)} \in \Omega$. Using (1.15) we define $\mu'_{(\Lambda)} = \mu_{(\Lambda)\eta_{L\backslash\Lambda}}$. We thus have

$$B_\Lambda(\xi_\Lambda) = W_{\Lambda,L\backslash\Lambda}(\xi_\Lambda \vee \eta_{L\backslash\Lambda}).$$

If $\Lambda \subset M$, $\xi_\Lambda \subset \Omega_\Lambda$, $\eta \in \Omega_{M\backslash\Lambda}$, we can write by (1.8)

$$\begin{aligned} B_M(\xi_\Lambda \vee \eta) &= W_{M,L\backslash M}(\xi_\Lambda \vee \eta \vee \eta_{L\backslash M}) \\ &= B'(\eta) + B''(\xi_\Lambda, \eta), \end{aligned}$$

where

$$\begin{aligned} B'(\eta) &= \sum_X {}^{*}\Phi(\eta \vee \eta_{L\backslash M}|X) = W_{M\backslash\Lambda, L\backslash M}(\eta \vee \eta_{L\backslash M}), \\ B''(\xi_\Lambda, \eta) &= \sum_X {}^{**}\Phi(\xi_\Lambda \vee \eta \vee \eta_{L\backslash M}|X). \end{aligned}$$

The sum $\sum^*$ extends over the finite sets $X \subset L\backslash\Lambda$ such that $X \cap (M\backslash\Lambda) \neq \emptyset$ and $X \cap (L\backslash M) \neq \emptyset$. The sum $\sum^{**}$ extends over the finite sets $X \subset L$ such that $X \cap \Lambda \neq \emptyset$ and $X \cap (L\backslash M) \neq \emptyset$. We now have to worry about the fact that $\xi_\Lambda \vee \eta \vee \eta_{L\backslash\Lambda}$ need not be defined, i.e. that there need be no $\zeta \in \Omega$ with restrictions $\xi_\Lambda, \eta, \eta_{L\backslash M}$ to $\Omega_\Lambda, \Omega_{M\backslash\Lambda}, \Omega_{L\backslash M}$. Given Λ, if M is sufficiently large this means that either $\xi_\Lambda \vee \eta$ or $\eta \vee \eta_{L\backslash M}$ is not defined (we use the fact that $\mathcal{F}$ is locally finite). Our conventions are thus respected in (1.18) by putting $e^{-W_{\Lambda,M\backslash\Lambda}(\xi_\Lambda \vee \eta)} = 0$ if $\xi_\Lambda \vee \eta$ is not defined, $e^{-B'(\eta)} = 0$ if $\eta \vee \eta_{L\backslash M}$ is not defined, and $\Phi(\xi_\Lambda \vee \eta \vee \eta_{L\backslash M}|X) = 0$ in B'' if $\xi_\Lambda \vee \eta \vee \eta_{L\backslash M}$ is not defined. The uniform convergence of (1.19) to the limit $\eta \mapsto \exp[-U_\Lambda(\xi_\Lambda) - W_{\Lambda,L\backslash\Lambda}(\xi_\Lambda \vee \eta)]$ is then easily checked. In conclusion a thermodynamic limit of $(\mu_{(\Lambda)\eta_{L\backslash\Lambda}})$ is necessarily a Gibbs state.

We now derive an important result from the above estimates. We have

$$\begin{aligned} \mu_{(M)\eta_{L\backslash M}}\{\xi_\Lambda \vee \eta\} = \left[Z'^{-1}_M e^{-U_{M\backslash\Lambda}(\eta) - B'(\eta)}\right] \\ \times \exp\left[-U_\Lambda(\xi_\Lambda) - W_{\Lambda,M\backslash\Lambda}(\xi_\Lambda \vee \eta) - B''(\xi_\Lambda, \eta)\right], \end{aligned} \tag{1.20}$$

where B'' is uniformly small for large M. We remark that by the definition (1.14), a Gibbs state σ is such that $\alpha_M\sigma$ is an average over $\eta_{L\backslash M}$ of the measures $\mu_{(M)\eta_{L\backslash M}}$. We can thus estimate the conditional probability with respect to $\sigma(d\xi)$ that $\xi|\Lambda = \xi_\Lambda$ when it is known that $\xi|(M\backslash\Lambda) = \eta$. According to (1.20), this is given by

$$\frac{(\alpha_M\sigma)\{\xi_\Lambda \vee \eta\}}{\sum_{\eta_\Lambda \in \Omega_\Lambda} (\alpha_M\sigma)\{\eta_\Lambda \vee \eta\}} \approx \frac{e^{-U_\Lambda(\xi_\Lambda) - W_{\Lambda,M\backslash\Lambda}(\xi_\Lambda \vee \eta)}}{\sum_{\eta_\Lambda \in \Omega_\Lambda} e^{-U_\Lambda(\eta_\Lambda) - W_{\Lambda,M\backslash\Lambda}(\eta_\Lambda \vee \eta)}},$$

where the error is uniformly as small as one likes for large M. From this follows that for a Gibbs state σ the conditional probability that $\xi|\Lambda = \xi_\Lambda$, when it is known that $\xi|(L\backslash\Lambda) = \eta$, is $\mu_{(\Lambda)\eta}\{\xi_\Lambda\}$. Conversely, if the conditional probability has this form, (1.14) clearly holds.

1.8 Theorem

A probability measure σ on Ω is a Gibbs state if and only if, for each finite $\Lambda \subset L$, the conditional probability that $\xi|\Lambda = \xi_\Lambda$, knowing that $\xi|(L\backslash\Lambda) = \eta$, is $\mu_{(\Lambda)\eta}\{\xi_\Lambda\}$ as given by (1.15).

Notice that we may thus take $\sigma_{L\backslash\Lambda} = \alpha_{L\backslash\Lambda}\sigma$ in (1.14). In particular a vague limit of Gibbs states is a Gibbs state. The set of Gibbs states is thus compact; it is clearly also convex.

1.9 Theorem

Let Φ be an interaction. The probability measures $\mu_{(\Lambda)}$ and $\mu_{(\Lambda)\eta}$ on Ω_Λ (Gibbs ensemble and Gibbs ensemble with boundary term) *are defined by* (1.11) *and* (1.15) *where $\eta \in \Omega_{L\backslash\Lambda}$ is the restriction to $L\backslash\Lambda$ of some element η^* of Ω (which may depend on Λ). Thermodynamic limits are defined by Proposition* 1.4.

(a) *Any thermodynamic limit of $(\mu_{(\Lambda)})$ is a Gibbs state.*
(b) *Any thermodynamic limit of $(\mu_{(\Lambda)_{\eta_{L\backslash\Lambda}}})$ is a Gibbs state.*
(c) *The closed convex hull of the Gibbs states obtained in* (b) *is the set K_Φ of all Gibbs states.*
(d) *$K_\Phi \neq \emptyset$; K_Φ is convex and compact; K_Φ is a Choquet simplex.*

(a) and (b) have been proved in Sections 1.6 and 1.7 respectively.

Because of (a) the set K_Φ of Gibbs states is not empty, and we have seen after Theorem 1.8 that K_Φ is convex and compact. Therefore the closed convex hull K of the Gibbs states obtained in (b) is contained in K_Φ. Suppose $K \neq K_\Phi$. There is then $A \in \mathcal{C}$ and $\sigma \in K_\Phi$ such that

$$\sigma(A) > \max_{\rho \in K} \rho(A), \tag{1.21}$$

and we may assume that $A = B \circ \alpha_\Lambda$, $B \in \mathcal{C}(\Omega_\Lambda)$ for some finite Λ. By (1.14) we have $\mu_{(M)_\eta}(B \circ \alpha_{\Lambda M}) \geqslant \sigma(A)$ for some $\eta \in \Omega_{L\backslash M}$ and all $M \supset \Lambda$. If ρ is a thermodynamic limit of $(\mu_{(M)_\eta})$, $\rho(A) \geqslant \sigma(A)$ in contradiction with (1.21). We have thus proved (c) and the first two parts of (d).

Let $\xi_\Lambda \in \Omega_\Lambda$ for a finite $\Lambda \subset L$. If $\sigma(d\xi)$ is any measure on Ω, let $\sigma_{(L\backslash\Lambda)\xi_\Lambda}$ be the measure on $\Omega_{L\backslash\Lambda}$ obtained by "setting $\xi|\Lambda = \xi_\Lambda$" (restrict σ to $\{\xi \in \Omega : \xi|\Lambda = \xi_\Lambda\}$, then take the image of the restriction by $\alpha_{L\backslash\Lambda}$). If σ is a Gibbs state, then by Theorem 1.8,

$$\sigma_{(L\backslash\Lambda)\xi_\Lambda}(d\eta) = \mu_{(\Lambda)\eta}\{\xi_\Lambda\}\sigma_{L\backslash\Lambda}(d\eta); \tag{1.22}$$

hence

$$e^{-U_\Lambda(\xi_\Lambda)-W_{\Lambda,L\setminus\Lambda}(\xi_\Lambda\vee\eta)}\sigma_{(L\setminus\Lambda)\eta_\Lambda}(d\eta) = e^{-U_\Lambda(\eta_\Lambda)-W_{\Lambda,L\setminus\Lambda}(\eta_\Lambda\vee\eta)}\sigma_{(L\setminus\Lambda)\xi_\Lambda}(d\eta) \tag{1.23}$$

for all $\xi_\Lambda, \eta_\Lambda \in \Omega_\Lambda$. Conversely if σ is a probability measure and (1.23) holds for all $\Lambda, \xi_\Lambda, \eta_\Lambda$, then (1.22) holds and σ is a Gibbs state. Consider now the closed linear subspace $\mathcal{G}$ of $\mathcal{C}^*$ consisting of the real measures σ such that (1.23) holds for all $\Lambda, \xi_\Lambda, \eta_\Lambda$. It is clear that if $\sigma \in \mathcal{G}$, also $|\sigma| \in \mathcal{G}$. From this it follows that K_Φ is a simplex (see Appendix A.5.5).

1.10 Algebra at infinity

Let σ be a probability measure on Ω. We let $\pi(A)$ be the class of A in $L^\infty(\Omega, \sigma)$. Define

$$\mathcal{B}_\sigma = \bigcap_{\Lambda \text{finite} \subset L} \text{closure}\left[\bigcup_{M \text{finite} \subset L\setminus\Lambda} \pi(\mathcal{C}_M)\right], \tag{1.24}$$

where the closure is taken with respect to the topology of $L^\infty(\Omega, \sigma)$ as weak dual of $L^1(\Omega, \sigma)$. Then $\mathcal{B}_\sigma$ will be called the *algebra at infinity* associated with σ. We shall now characterize the σ which have *trivial algebra at infinity* (i.e. $\mathcal{B}_\sigma$ consists of the almost everywhere constant functions) by the following *cluster property*.

(C) *For all $A \in \mathcal{C}$ there exists Λ finite $\subset L$ such that*

$$(B \in \mathcal{C}_M, \quad M \cap \Lambda = \emptyset) \Rightarrow |\sigma(AB) - \sigma(A)\sigma(B)| \leqslant \|B\|.$$

Suppose first that (C) holds. Then for all $A \in \mathcal{C}$ and $B_\infty \in \mathcal{B}_\sigma$,

$$|\sigma(AB_\infty) - \sigma(A)\sigma(B_\infty)| \leqslant \|B_\infty\|.$$

Replacing A by λA and letting $\lambda \to \infty$, we obtain

$$\sigma(AB_\infty) = \sigma(A)\sigma(B_\infty);$$

hence B_∞ is constant and $\mathcal{B}_\sigma$ is trivial.

Suppose now that (C) does not hold. There exist then $A \in \mathcal{C}$, and $B_{L\setminus\Lambda}$ for each finite $\Lambda \subset L$, such that

$$B_{L\setminus\Lambda} \subset \bigcup_{M \subset L\setminus\Lambda} \mathcal{C}_M, \quad \|B_{L\setminus\Lambda}\| = 1,$$

$$|\sigma(AB_{L\setminus\Lambda}) - \sigma(A)\sigma(B_{L\setminus\Lambda})| > 1.$$

Let B_∞ be a weak limit of the net $(\pi(B_{L\setminus\Lambda}))$ in $L^\infty(\Omega,\sigma)$; then $B_\infty \in \mathcal{B}_\sigma$ and we have

$$|\sigma(AB_\infty) - \sigma(A)\sigma(B_\infty)| \geqslant 1.$$

Therefore B_∞ cannot be constant: $\mathcal{B}_\sigma$ is not trivial.

We can now characterize the extremal points of the set K_Φ of Gibbs states (*pure Gibbs states*).

1.11 Theorem (characterization of pure Gibbs states)

Let $\sigma \in K_\Phi$; then the following conditions are equivalent;

(A) *σ is an extremal point of K_Φ.*
(B) *The algebra at infinity $\mathcal{B}_\sigma$ associated with σ is trivial.*
(C) *For all $A \in \mathcal{C}$ there exists Λ finite $\subset L$ such that*

$$(B \in \mathcal{C}_M, \quad M \cap \Lambda = \emptyset) \Rightarrow |\sigma(AB) - \sigma(A)\sigma(B)| \leqslant \|B\|.$$

We showed the equivalence of (B) and (C) in Section 1.10. We now prove $(A) \Leftrightarrow (B)$.

The non-extremality of σ in K_Φ is equivalent to the existence of $B_\infty \in L^\infty(\Omega,\sigma)$, $B_\infty \geqslant 0$, such that B_∞ is not a constant and $B_\infty\sigma$ is proportional to a Gibbs state. By (1.23) this is equivalent to

$$e^{-U_\Lambda(\xi_\Lambda) - W_{\Lambda,L\setminus\Lambda}(\xi_\Lambda \vee \eta)} B_\infty(\eta_\Lambda \vee \eta)\sigma_{(L\setminus\Lambda)\eta_\Lambda}(d\eta)$$
$$= e^{-U_\Lambda(\eta_\Lambda) - W_{\Lambda,L\setminus\Lambda}(\eta_\Lambda \vee \eta)} B_\infty(\xi_\Lambda \vee \eta)\sigma_{(L\setminus\Lambda)\xi_\Lambda}(d\eta)$$

for all Λ finite and $\xi_\Lambda, \eta_\Lambda \in \Omega_\Lambda$. But this means

$$B_\infty(\xi_\Lambda \vee \eta) = B_\infty(\eta_\Lambda \vee \eta)$$

(almost everywhere with respect to $(\alpha_{L\setminus\Lambda}\sigma)(d\eta)$) when $\xi_\Lambda \vee \eta, \eta_\Lambda \vee \eta$ are defined. Equivalently B_∞ is in the closure of $\cup_{M \subset L\setminus\Lambda}\pi(\mathcal{C}_M)$ for all Λ, and thus $B_\infty \in \mathcal{B}_\sigma$. The non-extremality of σ is thus equivalent to the existence of a non-trivial $B_\infty \in \mathcal{B}_\sigma$.

1.12 The operators $\mathcal{M}_\Lambda$

Given a finite $\Lambda \subset L$ we define a linear map $\mathcal{M}_\Lambda : \mathcal{C} \mapsto \mathcal{C}$ by

$$(\mathcal{M}_\Lambda A)(\xi_\Lambda \vee \eta) = \sum_{\eta_\Lambda \in \Omega_\Lambda} \mu_{(\Lambda)\eta}\{\eta_\Lambda\} A(\eta_\Lambda \vee \eta) \tag{1.25}$$

for all $\xi_\Lambda \vee \eta \in \Omega$, with $\xi_\Lambda \in \Omega_\Lambda$, $\eta \in \Omega_{L\setminus\Lambda}$.

Clearly $\|\mathcal{M}_\Lambda A\| \leqslant \|A\|$ and there is a function B on $\Omega_{L\backslash\Lambda}$ such that $(\mathcal{M}_\Lambda A)(\xi_\Lambda \vee \eta) = B(\eta)$. Therefore, if σ is a Gibbs state, (1.22) yields

$$\sigma(\mathcal{M}_\Lambda(A)) = (\alpha_{L\backslash\Lambda}\sigma)(B) = \int (\alpha_{L\backslash\Lambda}\sigma)(d\eta) \sum_{\eta_\Lambda \in \Omega_\Lambda} \mu_{(\Lambda)\eta}\{\eta_\Lambda\} A(\eta_\Lambda \vee \eta)$$

$$= \sum_{\eta_\Lambda \in \Omega_\Lambda} \int \sigma_{(L\backslash\Lambda)\eta_\Lambda}(d\eta) A(\eta_\Lambda \vee \eta) = \sigma(A). \tag{1.26}$$

(Conversely if the state σ is such that $\sigma(\mathcal{M}_\Lambda A) = \sigma(A)$ for all Λ and A, then σ is a Gibbs state.)

Suppose that, for each A, $\mathcal{M}_\Lambda A$ tends in $\mathcal{C}$ to a constant c when $\Lambda \to L$ (i.e. given $\epsilon > 0$ there exists Δ finite such that $\|\mathcal{M}_\Lambda(A) - c\| < \epsilon$ if $\Lambda \supset \Delta$). Let then $\sigma, \sigma' \in K_{\Phi'}$; we have

$$\sigma(A) = \lim_{\Lambda\to L} \sigma(\mathcal{M}_\Lambda(A)) = \lim_{\Lambda\to L} \sigma'(\mathcal{M}_\Lambda(A)) = \sigma'(A)$$

and therefore K_Φ consists of a single point.

Suppose now that $\mathcal{M}_\Lambda A$ does not tend to a constant limit. We assume, as we may, that $A = B \circ \alpha_\Lambda \in \mathcal{C}_\Lambda$ for some Λ. We can find sequences (M_n), (M'_n) and $\xi_n, \xi'_n \in \Omega$ such that $M_n \to L$, $M'_n \to L$ and

$$\lim_{n\to\infty} (\mathcal{M}_{M_n}(A))(\xi_n) \neq \lim_{n\to\infty} (\mathcal{M}_{M'_n}(A))(\xi'_n).$$

But this means

$$\lim_{n\to\infty} \mu_{(M_n)\eta_n}(B \circ \alpha_{\Lambda M_n}) \neq \lim_{n\to\infty} \mu_{(M'_n)\eta'_n}(B \circ \alpha_{\Lambda M_n}),$$

where $\eta_n = \xi_n|(L\backslash M_n)$, $\eta'_n = \xi'_n|(L\backslash M'_n)$. Therefore K_Φ consists of more than one point.

1.13 Theorem (characterization of unique Gibbs states)

Let $\sigma \in K_\Phi$; then we have the implications (A′) $\Leftrightarrow$ (B′) $\Rightarrow$ (C′).

(A′) $K_\Phi = \{\rho\}$.
(B′) *For all $A \in \mathcal{C}$, $\mathcal{M}_\Lambda A$ tends in $\mathcal{C}$ to a constant when $\Lambda \to L$.*
(C′) *For all $A \in \mathcal{C}$ there exists Λ finite $\subset L$ such that*

$$(0 \leqslant B \in \mathcal{C}_M, \quad M \cap \Lambda = \emptyset) \Rightarrow |\sigma(AB) - \sigma(A)\sigma(B)| \leqslant \sigma(B).$$

Furthermore (C′) $\Rightarrow$ (B′) *if* supp $\sigma = \Omega$.

Notice that (C′) is a *cluster property* stronger than (C) of Section 1.10.

We have shown in Section 1.12 that (A′) ⇔ (B′). If (B′) holds and A is given we can choose Λ such that

$$|\mathfrak{M}_\Lambda A - \sigma(A)| \leqslant 1;$$

hence if $B \in \mathcal{C}_M$, $M \cap \Lambda = \emptyset$, then

$$\sigma(AB) = \sigma(\mathfrak{M}_\Lambda(AB)) = \sigma((\mathfrak{M}_\Lambda A) \cdot B);$$

and if $B \geqslant 0$,

$$|\sigma(AB) - \sigma(A)\sigma(B)| = |\sigma((\mathfrak{M}_\Lambda A - \sigma(A))B)| \leqslant \sigma(B).$$

Thus (B′) ⇒ (C′).

We prove now that (C′) ⇒ (B′) if supp $\sigma = \Omega$. Suppose that (B′) does not hold. Then there exist A and sequences (Λ_n), (ξ_n) such that $\Lambda_n \to L$ and $(\mathfrak{M}_{\Lambda_n} A)(\xi_n) \to c \neq \sigma(A)$. By changing A we can assume that $c - \sigma(A) = 4$. By continuity of $\mathfrak{M}_{\Lambda_n} A$ we can find M_n such that $M_n \cap \Lambda_n = \emptyset$ and $|(\mathfrak{M}_{\Lambda_n} A)(\eta) - (\mathfrak{M}_{\Lambda_n} A)(\xi_n)| < 1$ if $\eta|M_n = \xi_n|M_n$. Let B_n be the characteristic function of $\{\eta \in \Omega : \eta|M_n = \xi_n|M_n\}$; then for sufficiently large n

$$(\mathfrak{M}_{\Lambda_n} A)(\eta) > \sigma(A) + 2 \quad \text{if} \quad B_n(\eta) \neq 0.$$

Therefore

$$\sigma(AB_n) = \sigma(\mathfrak{M}_{\Lambda_n}(AB_n)) = \sigma((\mathfrak{M}_{\Lambda_n} A) \cdot B_n) \geqslant (\sigma(A) + 2)\sigma(B_n);$$

hence

$$|\sigma(AB_n) - \sigma(A)\sigma(B_n)| \geqslant 2\sigma(B_n).$$

This contradicts (C′) because $\sigma(B_n) > 0$ by our assumption supp $\sigma = \Omega$.

1.14 Remark

Suppose that Ω has the following property.

(D*) *For all* $\xi, \eta \in \Omega$ *and* Λ *finite* $\subset L$, *there exists a finite* $M \subset L$, *and* $\zeta \in \Omega$ *such that*

$$\zeta|\Lambda = \xi|\Lambda, \quad \zeta|(L\backslash M) = \eta|(L\backslash M)$$

(in other words, for all $\eta \in \Omega$*, the set* $\sum_\eta = \{\zeta \in \Omega : \exists M$ *finite and* $\zeta|(L\backslash M) = \eta|(L\backslash M)\}$ *is dense in* Ω*).*

Then, for every Gibbs state σ, supp $\sigma = \Omega$ (this follows readily from the definition (1.14) of Gibbs states). The conditions (A′), (B′), (C′) of Theorem 1.13 are then all equivalent.

Notes

The important notion of Gibbs state was introduced by Dobrushin [1], [2], [3] (and rediscovered by Lanford and Ruelle [1]). Gibbs states are probability measures for which certain conditional probabilities are prescribed (Theorem 1.8). Equivalently they are probability measures satisfying a certain set of equations (14) sometimes called the DLR equations. In this chapter, we largely follow Dobrushin, with a slightly more general framework and some additions: the simplex property and the notion of algebra at infinity (Lanford and Ruelle [1]), the $\mathfrak{M}_\Lambda$ operators (Ledrappier [1]).

Exercises

1. If $\Omega = \prod_{x \in L} \Omega_x$ ($\mathcal{F}$ empty) and $\Phi | \Omega_\Lambda = 0$ for $|\Lambda| > 1$, there is only one Gibbs state: $\sigma = \prod_{x \in L} \sigma_{(\{x\})}$.
2. Show that the set of thermodynamic limits of $(\mu_{(\Lambda)\eta_{L \setminus \Lambda}})$ – see Theorem 1.9(b) – is closed, and contains all the extremal points of K_Φ. [The second point, first noted by Georgii [2], follows easily from Milman's theorem: see Appendix A.3.5.]
3. Two distinct pure Gibbs states are disjoint [i.e. mutually singular] measures on Ω. (See Appendix A.5.5.)
4. Let Φ, Ψ be interactions and $\sigma \in K_\Phi$.

   (a) For finite $\Lambda \subset L$, and $\xi \in \Omega_\Lambda$, define

   $$\mu^*_{(\Lambda)}\{\xi\} = Z^{*-1} \exp\left[-U^\Psi_\Lambda(\xi)\right] \cdot (\alpha_\Lambda \sigma)\{\xi\},$$
   $$Z^* = (\alpha_\Lambda \sigma)\left(\exp\left[-U^\Psi_\Lambda\right]\right).$$

   Prove that any thermodynamic limit of $(\mu^*_{(\Lambda)})$ belongs to $K_{\Phi+\Psi}$.

   (b) Using (a), show that when $\Lambda \to L$, any limit of

   $$Z^{*-1} \exp\left[-U^\Psi_\Lambda(\zeta \,|\, \Lambda)\right] \cdot \sigma(d\zeta)$$

   belongs to $K_{\Phi+\Psi}$.

   (c) Find analogues of (a) and (b) where U_Λ is replaced by $U_\Lambda + W_{\Lambda, L \setminus \Lambda}$. [Caution: the denominator Z^* must be different from zero.]

2

Gibbs states: complements

In this chapter we study how interactions and Gibbs states transform under maps.

2.1 Morphisms of lattice systems

We shall call the object $(L, (\Omega_x)_{x \in L}, (\bar{\Omega}_\Lambda)_{\Lambda \in \mathfrak{F}})$ introduced in Chapter 1 a *lattice system*. To this lattice system we have associated a configuration space Ω defined in Section 1.1. Let Ω' be similarly the configuration space of the lattice system $(L', (\Omega'_{x'})_{x' \in L'}, (\bar{\Omega}'_{\Lambda'})_{\Lambda' \in \mathfrak{F}'})$. Suppose now that the family $(F_x)_{x \in L}$ has the following properties.

(M1) *F_x is a map $\Omega'_{M(x)} \mapsto \Omega_x$ where $M(x)$ is a finite subset of L'.*

(M2) *The family $(M(x))_{x \in L}$ is locally finite (i.e., $\{x : x' \in M(x)\}$ is finite for each $x' \in L'$).*

(M3) *If $\xi' \in \Omega'_{\cup\{M(x):x \in X\}}$ then $(F_x(\xi'|M(x)))_{x \in X}$ is an element $FX\xi'$ of ΩX whenever $X \subset L$.*

It suffices to check (M3) for $X \in \mathfrak{F}$.

A continuous map $F : \Omega' \mapsto \Omega$ is defined by

$$(F\xi')_x = F_x(\xi'|M(x)).$$

Let $\sum_\xi \subset \Omega$ consist of those η such that $\eta_x = \xi_x$ except for finitely many $x \in L$, and similarly for $\sum'_{\xi'} \subset \Omega'$. Then F maps $\sum'_\xi$ into $\sum_{F\xi'}$. We say that F is a *morphism*, from $(L', (\Omega'_{x'})_{x' \in L'}, (\bar{\Omega}'_{\Lambda'})_{\Lambda' \in \mathfrak{F}'})$ to $(L, (\Omega_x)_{x \in L}, (\bar{\Omega}_\Lambda)_{\Lambda \in \mathfrak{F}})$, if it satisfies

(M4) *F restricted to $\sum'_{\xi'}$ is a bijection to $\sum_{F\xi'}$ for each $\xi' \in \Omega'$.*

Notice that distinct families $(F_x)_{x \in L}$ may define the same map F, and therefore the same morphism. It is readily verified that the identity map of

Ω is a morphism (identity morphism) and that composition of two morphisms yields another morphism (Exercise 1). Suppose that F' is a morphism from $(L, (\Omega_x)_{x\in L}, (\bar{\Omega}_\Lambda)_{\Lambda\in\mathfrak{F}})$ to $(L', (\Omega'_{x'})_{x'\in L'}, (\bar{\Omega}'_{\Lambda'})_{\Lambda'\in\mathfrak{F}'})$, and that FF' and $F'F$ are the identity maps on Ω and Ω' respectively; then F is called an *isomorphism*.

2.2 Example

Suppose that the lattice system $(L', (\Omega'_{x'})_{x'\in L'}, (\bar{\Omega}_{\Lambda'})_{\Lambda'\in\mathfrak{F}'})$ is given. Let $(M(x))_{x\in L}$ be a partition of L' into finite subsets. Define $\Omega_x = \Omega'_{M(x)}$ and $F_x : \Omega'_{M(x)} \mapsto \Omega_x$ be the identity. We take $\Lambda \in \mathfrak{F}$ if there exists $\Lambda' \in \mathfrak{F}'$ such that $\Lambda' \cap M(x) \neq \emptyset$ for all $x \in \Lambda$, we may then define $\bar{\Omega}_\Lambda = \{(\xi'|M(x))_{x\in\Lambda} : \xi' \in \Omega'_{\cup\{M(x):x\in\Lambda\}}\}$. It is easily seen that $(F_x)_{x\in L}$ defines a morphism F from $(L', (\Omega'_{x'})_{x'\in L'}, (\bar{\Omega}'_{\Lambda'})_{\Lambda'\in\mathfrak{F}'})$ to $(L, (\Omega_x)_{x\in L}, (\bar{\Omega}_\Lambda)_{\Lambda\in\mathfrak{F}})$, and in fact an isomorphism.

2.3 The interaction $F^*\Phi$

Let F be a morphism defined by a family $(F_x)_{x\in L}$ as in Section 2.1. Given an interaction Φ for $(L, (\Omega_x)_{x\in L}, (\bar{\Omega}_\Lambda)_{\Lambda\in\mathfrak{F}})$, we introduce an interaction $F^*\Phi$ for $(L', (\Omega'_{x'})_{x'\in L'}, (\bar{\Omega}'_{\Lambda'})_{\Lambda'\in\mathfrak{F}'})$ by

$$(F^*\Phi)(\xi') = \sum_{X:\cup\{M(x):x\in X\}=X'} \Phi(F_x\xi') \qquad \text{if} \quad \xi' \in \Omega_{X'}.$$

The sum is finite because of (M2). Furthermore, using (M3),

$$\begin{aligned}
\|F^*\Phi\|_{x'} &= \sum_{X'\ni x'} \sup_{\xi'\in\Omega'_{x'}} |(F^*\Phi)(\xi')| \\
&\leqslant \sum_{\substack{X'\ni x' \\ X:\cup\{M(x):x\in X\}=X'}} \sup_{\xi'\in\Omega'_{X'}} |\Phi(F_X\xi')| \\
&= \sum_{X:\cup\{M(x):x\in X\}\ni x'} \sup_{\xi'\in\Omega'_{X'}} |\Phi(F_X\xi')| \\
&\leqslant \sum_{X:X\cap\{x:x'\in M(x)\}\neq\emptyset} \sup_{\xi\in\Omega_X} |\Phi(\xi)| \\
&\leqslant \sum_{x:x'\in M(x)} \|\Phi\|_x.
\end{aligned}$$

Notice that $F^*\Phi$ depends on $(F_x)_{x\in L}$ and not only on the morphism F.

2.4 Lemma

The conditional probabilities $\mu^{F^*\Phi}_{(\Lambda')\xi'_{L\backslash\Lambda'}}\{\xi'_{\Lambda'}\}$ *depend only on the morphism* F, *and not otherwise on* $(F_x)_{x\in L}$.

We have

$$\mu_{F^*\Phi(\Lambda')\xi'_{L'\backslash\Lambda'}}\{\xi'_{\Lambda'}\} = \frac{\exp[-U_{\Lambda'}(\xi'_{\Lambda'}) - W_{\Lambda',L'\backslash\Lambda'}(\xi'_{\Lambda'} \vee \xi'_{L'\backslash\Lambda'})]}{\sum_{\eta'_{\Lambda'}\in\Omega'_{\Lambda'}} \exp[-U_{\Lambda'}(\eta'_{\Lambda'}) - W_{\Lambda',L'\backslash\Lambda'}(\eta'_{\Lambda'} \vee \xi'_{L'\backslash\Lambda'})]}, \tag{2.1}$$

where U, W are computed from the interaction $F^*\Phi$. Thus

$$\begin{aligned}
&\exp[-U_{\Lambda'}(\eta'_{\Lambda'}) - W_{\Lambda',L'\backslash\Lambda'}(\eta'_{\Lambda'} \vee \xi'_{L'\backslash\Lambda'})] \\
&= \exp\left[-\sum_{X':X'\cap\Lambda'\neq\emptyset}\ \sum_{X:\cup\{M(x):x\in X\}=X'} \Phi(F_X((\eta'_{\Lambda'} \vee \xi'_{L'\backslash\Lambda'})|X'))\right] \\
&= \exp\left[-\sum_{X:\cup\{M(x):x\in X\}\cap\Lambda'\neq\emptyset} \Phi(F(\eta'_{\Lambda'} \vee \xi'_{L'\backslash\Lambda'})|X))\right] \\
&= \prod_{X:\cup\{M(x):x\in X\}\cap\Lambda'\neq\emptyset} \exp\left[-\Phi(F(\eta'_{\Lambda'} \vee \xi'_{L'\backslash\Lambda'})|X)\right].
\end{aligned} \tag{2.2}$$

Another choice of $(F_x)_{x\in L}$ would give the same expression apart from a change in the sets $M(x)$; (2.1) would however remain the same due to cancellations between numerator and denominator.

2.5 Proposition

If σ' *is a Gibbs state on* Ω' *for the interaction* $F^*\Phi$, *then* $F\sigma'$ *is a Gibbs state on* Ω *for the interaction* Φ.

Let $\xi' \in \Omega'$, $\xi = F\xi' \in \Omega$, and let Λ, Λ' be finite subsets of L, L' respectively. We write

$$\xi_\Lambda = \xi|\Lambda, \quad \xi_{L\backslash\Lambda} = \xi|(L\backslash\Lambda), \quad \xi'_{L'\backslash\Lambda'} = \xi'|(L'\backslash\Lambda').$$

In view of (M4) there is, for given Λ and sufficiently large Λ', a bijection f from

$$A = \{\eta_\Lambda \in \Omega_\Lambda : \eta_\Lambda \vee \xi_{L\backslash\Lambda} \text{ is defined}\}$$

to

$$\{\eta'_{\Lambda'} \in \Omega'_{\Lambda'} : \eta'_{\Lambda'} \vee \xi'_{L'\backslash\Lambda'} \text{ is defined and } F(\eta'_{\Lambda'} \vee \xi'_{L'\backslash\Lambda'})|(L \setminus \Lambda) = \xi_{L\backslash\Lambda}\}$$

such that

$$F((f\eta_\Lambda) \vee \xi'_{L'\setminus\Lambda'}) = \eta_\Lambda \vee \xi_{L\setminus\Lambda}. \tag{2.3}$$

Let now P be the conditional probability with respect to $\sigma'(d\eta')$ that $(F\eta')|\Lambda = \xi_\Lambda$, knowing that $\eta'|(L'\setminus\Lambda') = \xi'_{L'\setminus\Lambda'}$ and that $(F\eta')|(L\setminus\Lambda) = \xi_{L\setminus\Lambda'}$. We have, since σ' is a Gibbs state,

$$P = \frac{\mu^{F^*\Phi}_{(\Lambda')\xi'_{L'\setminus\Lambda'}}\{f\xi_\Lambda\}}{\sum_{\eta_\Lambda\in A}\mu^{F^*\Phi}_{(\Lambda')\xi'_{L'\setminus\Lambda'}}\{f\eta_\Lambda\}}$$

$$= \frac{\exp\left[-U_{\Lambda'}(f\xi_\Lambda) - W_{\Lambda',L'\setminus\Lambda'}((f\xi_\Lambda)\vee\xi'_{L'\setminus\Lambda'})\right]}{\sum_{\eta_\Lambda\in A}\exp\left[-U_{\Lambda'}(f\eta_\Lambda) - W_{\Lambda',L'\setminus\Lambda'}((f\eta_\Lambda)\vee\xi'_{L'\setminus\Lambda'})\right]}.$$

Using (2.2) and (2.3), this is

$$P = \frac{\prod_{X:\cup\{M(x):x\in X\}\cap\Lambda'\neq\emptyset}\exp[-\Phi(\xi_\Lambda\vee\xi_{L\setminus\Lambda}|X)]}{\sum_{\eta_\Lambda\in A}\prod_{X:\cup\{M(x):x\in X\}\cap\Lambda'\neq\emptyset}\exp[-\Phi(\eta_\Lambda\vee\xi_{L\setminus\Lambda}|X)]};$$

the factors with $X\cap\Lambda = \emptyset$ are the same in the numerator and denominator. Therefore

$$P = \frac{\exp[-\sum_{X:X\cap\Lambda\neq\emptyset}\Phi(\xi_\Lambda\vee\xi_{L\setminus\Lambda}|X)]}{\sum_{\eta_\Lambda\in\Omega_\Lambda}\exp[-\sum_{X:X\cap\Lambda\neq\emptyset}\Phi(\eta_\Lambda\vee\xi_{L\setminus\Lambda}|X)]} = \mu^\Phi_{(\Lambda)\xi_{L\setminus\Lambda}}\{\xi_\Lambda\}.$$

This does not depend on $\xi'_{L'\setminus\Lambda'}$, and therefore the conditional probability with respect to $\sigma'(d\eta')$ that $(F\eta')|\Lambda = \xi_\Lambda$, knowing that $(F\eta')|(L\setminus\Lambda) = \xi_{L\setminus\Lambda}$, is again $\mu^\Phi_{(\Lambda)\xi_{L\setminus\Lambda}}\{\xi_\Lambda\}$. This is then also the conditional probability with respect to $(F\sigma')(d\eta)$ that $\eta|\Lambda = \xi_\Lambda$, knowing that $\eta|(L\setminus\Lambda) = \xi_{L\setminus\Lambda}$. Therefore $F\sigma'$ is a Gibbs state for the interaction Φ.

2.6 Remarks

If $F\circ F'$ is the morphism obtained by composing two morphisms F, F', it is easily seen (Exercise 1) that

$$(F\circ F')^* = F'^*F^* \tag{2.4}$$

when $F\circ F'$ is defined by a suitable family $(\tilde{F}_x)_{x\in L}$.

If I is the identity morphism of $(L, (\Omega_x)_{x\in L}, (\bar{\Omega}_\Lambda)_{\Lambda\in\mathfrak{F}})$ and Φ is an interaction, Lemma 2.4 shows that Φ and $I^*\Phi$ define the same conditional probabilities:

$$\mu^{I^*\Phi}_{(\Lambda)\xi_{L\setminus\Lambda}}\{\xi_\Lambda\} = \mu^{\Phi}_{(\Lambda)\xi_{L\setminus\Lambda}}\{\xi_\Lambda\} \tag{2.5}$$

and have therefore the same Gibbs states.

If $F : \Omega' \mapsto \Omega$ is an isomorphism, F is a bijection of the set of Gibbs states for $F^*\Phi$ on Ω' to the set of Gibbs states for Φ on Ω. This follows from Proposition 2.5 and from (2.4), (2.5).

2.7 Systems of conditional probabilities

For every finite $\Lambda \subset L$, we write

$$\Omega^*_{L\setminus\Lambda} = \{\xi : (\exists \xi^* \in \Omega)\xi = \xi^*|(L\setminus\Lambda)\}.$$

This set is closed in $\Omega_{L\setminus\Lambda}$ (see Section 1.5). The conditional probabilities $\mu_{(\Lambda)\xi_{L\setminus\Lambda}}\{\xi_\Lambda\}$ for all allowed interactions Φ (i.e. $\|\Phi\|_x < +\infty$ for all x – see Section 1.2) satisfy the following conditions.

(a) *If* $\xi_{L\setminus\Lambda} \in \Omega^*_{L\setminus\Lambda}$ *and* $\xi_\Lambda \in \Omega_\Lambda$, *then* $\mu_{(\Lambda)\xi_{L\setminus\Lambda}}\{\xi_\Lambda\} \geqslant 0$ *and* $\sum_{\eta_\Lambda\in\Omega_\Lambda} \mu_{(\Lambda)\xi_{L\setminus\Lambda}}\{\eta_\Lambda\} = 1$. *Furthermore* $\mu_{(\Lambda)\xi_{L\setminus\Lambda}}\{\xi_\Lambda\} > 0$ *if and only if* $\xi_\Lambda \vee \xi_{L\setminus\Lambda} \in \Omega$.

(b) *If* $\xi_\Lambda \in \Omega_\Lambda$, *the real function* $\xi_{L\setminus\Lambda} \mapsto \mu_{(\Lambda)\xi_{L\setminus\Lambda}}\{\xi_\Lambda\}$ *on* $\Omega^*_{L\setminus\Lambda}$ *is continuous.*

(c) *Let* $\Lambda \subset M$, $\xi_\Lambda \in \Omega_\Lambda$, $\xi_{M\setminus\Lambda} \in \Omega_{L\setminus\Lambda}$, $\xi_{L\setminus M} \in \Omega_{L\setminus M}$, *and* $\xi_\Lambda \vee \xi_{M\setminus\Lambda} \vee \xi_{L\setminus M} \in \Omega$; *then*

$$\mu_{(\Lambda)\xi_{M\setminus\Lambda}\vee\xi_{L\setminus M}}\{\xi_\Lambda\} \times \sum_{\eta\in\Omega_\Lambda} \mu_{(M)\xi_{L\setminus M}}\{\eta \vee \xi_{M\setminus\Lambda}\}$$
$$= \mu_{(M)\xi_{L\setminus M}}\{\xi_\Lambda \vee \xi_{M\setminus\Lambda}\}.$$

(a) and (b) follow from Section 1.5; (c) agrees with the interpretation of the $\mu_{(\Lambda)\xi_{L\setminus\Lambda}}\{\xi_\Lambda\}$ as conditional probabilities and can be checked by direct computation.

A family $(\mu_{(\Lambda)\xi_{L\setminus\Lambda}})$ satisfying (a), (b), (c) will be called a *system of conditional probabilities*. The main part of the theory developed till now with interactions applies with trivial modifications to systems of conditional probabilities.

The definition of Gibbs states (Section 1.5) is unchanged. We now want to show that any thermodynamic limit of the $\mu_{(\Lambda)\xi_{L\setminus\Lambda}}$ (with $\xi_{L\setminus\Lambda} \in \Omega^*_{L\setminus\Lambda}$) is a

Gibbs state. Let $\Lambda \subset M \subset N \subset L$. The conditional probability with respect to $\alpha_{MN}\mu_{(N)\xi_{L\setminus N}}$ that $\eta|\Lambda = \xi_\Lambda$, knowing that $\eta|(M\setminus\Lambda) = \xi_{M\setminus\Lambda}$, is

$$P = \frac{\alpha_{MN}\mu_{(N)\xi_{L\setminus N}}\{\xi_\Lambda \vee \xi_{M\setminus\Lambda}\}}{\sum\limits_{\eta_\Lambda\in\Omega_\Lambda} \alpha_{MN}\mu_{(N)\xi_{L\setminus N}}\{\eta_\Lambda \vee \xi_{M\setminus\Lambda}\}}$$

$$= \frac{\sum\limits_{\eta_{N\setminus M}} \mu_{(N)\xi_{L\setminus N}}\{\xi_\Lambda \vee \xi_{M\setminus\Lambda} \vee \eta_{N\setminus M}\}}{\sum\limits_{\eta_\Lambda}\sum\limits_{\eta_{N\setminus M}} \mu_{(N)\xi_{L\setminus N}}\{\eta_\Lambda \vee \xi_{M\setminus\Lambda} \vee \eta_{N\setminus M}\}}.$$

If this expression is ambiguous we can define it in an arbitrary manner. Using (c) we get

$$P = \sum_{\eta_{N\setminus M}} \mu_{((N\setminus M)\cup\Lambda)\xi_{M\setminus\Lambda}\vee\xi_{L\setminus N}}\{\xi_\Lambda \vee \eta_{N\setminus M}\}$$

$$= \sum_{\eta_{N\setminus M}} \mu_{(\Lambda)\xi_{M\setminus\Lambda}\vee\eta_{N\setminus M}\vee\xi_{L\setminus N}}\{\xi_\Lambda\}$$

$$\times \sum_{\eta_\Lambda} \mu_{((N\setminus M)\cup\Lambda)\xi_{M\setminus\Lambda}\vee\xi_{L\setminus N}}\{\eta_\Lambda \vee \eta_{N\setminus M}\}.$$

Therefore, using (a), P is the average over $\eta_{N\setminus M}$, with respect to some weight, of

$$\mu_{(\Lambda)\xi_{M\setminus\Lambda}\vee\eta_{N\setminus M}\vee\xi_{L\setminus N}}\{\xi_\Lambda\}.$$

In view of the continuity property (b), the oscillation of this quantity with respect to $\eta_{N\setminus M} \vee \xi_{L\setminus N}$ tends to zero uniformly when $M \to \infty$.

From this it follows that a thermodynamic limit of $(\mu_{(\Lambda)\eta_{L\setminus\Lambda}})$ is a Gibbs state.

2.8 Properties of Gibbs states

The proofs of the results of Chapter 1, from Theorem 1.8 on, remain applicable with obvious changes in the present situation. More explicitly, Theorem 1.8 and parts (b), (c), (d) of Theorem 1.9 hold for systems of conditional probabilities. The characterization of pure Gibbs states (Theorem 1.11) and of unique Gibbs states (Theorem 1.13, Remark 1.14) remain valid.

Let a morphism F be defined as in Section 2.1, and $(\mu_{(\Lambda)\xi_{L\setminus\Lambda}})$ be a system of conditional probabilities for $(L, (\Omega_x)_{x\in L}, (\bar\Omega_\Lambda)_{\Lambda\in\mathfrak{F}})$. We now introduce a system $(F^*\mu_{(\Lambda')\xi'_{L'\setminus\Lambda'}})$ on $(L', (\Omega'_{x'})_{x'\in L'}, (\bar\Omega'_{\Lambda'})_{\Lambda'\in\mathfrak{F}'})$.

For finite $\Lambda' \subset L'$ and $\xi'_{L'\setminus\Lambda'} \in \Omega^*_{L'\setminus\Lambda'}$, let $\Lambda = \{x \in L : M(x) \cap \Lambda' \neq \emptyset\}$. Then, by the properties of a morphism,

$$F(\eta'_{\Lambda'} \vee \xi'_{L'\setminus\Lambda'}) = F_\Lambda(\eta'_{\Lambda'} \vee \xi'_{L'\setminus\Lambda'}) \vee \xi_{L\setminus\Lambda}$$

for some fixed $\xi_{L\backslash\Lambda} \in \Omega^*_{L\backslash\Lambda}$, whenever $\eta'_{\Lambda'} \vee \xi'_{L'\backslash\Lambda'} \in \Omega'$. We define

$$F^*\mu_{(\Lambda')\xi'_{L'\backslash\Lambda'}}\{\xi'_{\Lambda'}\} = \frac{\mu_{(\Lambda)\xi_{L\backslash\Lambda}}\{F_\Lambda(\xi'_{\Lambda'} \vee \xi'_{L'\backslash\Lambda'})\}}{\sum_{\eta'_{\Lambda'}\in\Omega'_{\Lambda'}} \mu_{(\Lambda)\xi_{L\backslash\Lambda}}\{F_\Lambda(\eta'_{\Lambda'} \vee \xi'_{L'\backslash\Lambda'})\}}.$$

The family $(F^*\mu_{(\Lambda')\xi'_{L'\backslash\Lambda'}})$ is a system of conditional probabilities on $(L', (\Omega'_{x'})_{x'\in L'}, (\bar{\Omega}'_{\Lambda'})_{\Lambda'\in\mathfrak{F}'})$. Notice that, by (2.1) and (2.2), if Φ is an interaction,

$$F^*\mu^{\Phi}_{(\Lambda')\xi'_{L'\backslash\Lambda'}} = \mu^{F^*\Phi}_{(\Lambda')\xi'_{L'\backslash\Lambda'}}.$$

It is not hard to see that Proposition 2.5 and Remarks 2.6 apply to systems of conditional probabilities (consider $(F^*\mu_{(\Lambda')\xi'_{L'\backslash\Lambda'}})$ instead of $F^*\Phi$).

2.9 Remark

The interest in considering systems of conditional probabilities rather than interactions is on the one hand that they are a priori more general, and on the other that they behave better under morphisms. Indeed, if F is a morphism, F^* is uniquely defined on systems of conditional probabilities, but not on interactions. Furthermore, on systems of conditional probabilities, morphisms act functorially in the sense that $(F \circ F')^* = F'^* F^*$, and that I^* is the identity if I is the identity morphism.

Systems of conditional probabilities are however not always the natural objects to consider*, and it is always convenient to have interactions at one's disposal.

Notes

This chapter is of a somewhat formal nature; it is an attempt at removing the arbitrariness present in the choice of L, $(\Omega_x)_{x\in L}$, $(\bar{\Omega}_\Lambda)_{\Lambda\in\mathfrak{F}}$ and the interactions Φ of Chapter 1. This is done by introducing an isomorphism notion between "lattice systems" $\big(L, (\Omega_x)_{x\in L}, (\bar{\Omega}_\Lambda)_{\Lambda\in\mathfrak{F}}\big)$, and by associating with each interaction a "system of conditional probabilities." One can define not only isomorphisms of lattice systems but – more interestingly – morphisms. These morphisms transport systems of conditional probabilities contravariantly, and Gibbs states

* This is because they do not permit the definition of the pressure P when there is invariance under translations (see Chapters 3, 4).

covariantly. Morphisms can be used to formalize some procedures commonly used in statistical mechanics.*

For the notion of system of conditional probabilities, see Sullivan [1]; the rest of this chapter is largely original.

Exercises

1. Let F, F' be morphisms

$$F : (L', (\Omega'_{x'})_{x' \in L'}, (\bar{\Omega}'_{\Lambda'})_{\Lambda' \in \mathfrak{F}'}) \mapsto (L, (\Omega_x)_{x \in L}, (\bar{\Omega}_\Lambda)_{\Lambda \in \mathfrak{F}}),$$
$$F' : (L'', (\Omega''_{x''})_{x'' \in L''}, (\bar{\Omega}''_{\Lambda''})_{\Lambda'' \in \mathfrak{F}''}) \mapsto (L', (\Omega'_{x'})_{x' \in L'}, (\bar{\Omega}'_{\Lambda'})_{\Lambda' \in \mathfrak{F}'})$$

   defined by families $(F_x)_{x \in L}, (F'_{x'})_{x' \in L'}$ with $F_x : \Omega'_{M(x)} \mapsto \Omega_x$ and $F'_{x'} : \Omega''_{M'(x')} \mapsto \Omega'_{x'}$. Let $\tilde{M}(x) = \cup\{M'(x') : x' \in M(x)\}$ and $\tilde{F}_x : \Omega''_{\tilde{M}(x)} \mapsto \Omega_x$ be such that $\tilde{F}(\xi'') = F_x(F'_{M(x)}\xi'')$.

   Check that the map $\tilde{F} = F \circ F'$ is a morphism defined by the family $(\tilde{F}_x)_{x \in L}$. Check that $\tilde{F}^*\Phi = F'^* F^* \Phi$ where F^*, F'^*, $\tilde{F}^*$ are defined by the families (F_x), $(F'_{x'})$, $(\tilde{F}_x)$ respectively.

2. If we suppress the condition that $\mu_{(\Lambda)\xi_{L\backslash\Lambda}}\{\xi_\Lambda\} > 0$ when $\xi_\Lambda \vee \xi_{L\backslash\Lambda} \in \Omega$ in Section 2.7(a), we obtain a more general notion of systems of conditional probabilities. Check that the properties listed in the first paragraph of Section 2.8 (except Remark 1.14) remain valid.

3. Let a system of conditional probabilities $(\mu_{(\Lambda)\xi_{L\backslash\Lambda}})$ be given. Whenever $\xi, \eta \in \Omega$ and there exists Λ finite such that $\xi|(L\backslash\Lambda) = \eta|(L\backslash\Lambda)$, define

$$V(\xi, \eta) = \log \mu_{(\Lambda)\xi|(L\backslash\Lambda)}\{\xi|\Lambda\} - \log \mu_{(\Lambda)\xi|(L\backslash\Lambda)}\{\eta|\Lambda\}. \qquad (*)$$

   Check that this definition does not depend on the choice of Λ, and that the following properties hold:

   (a) V is a real function on $\{(\xi, \eta) \in \Omega \times \Omega : \xi|(L\backslash\Lambda) = \eta|(L\backslash\Lambda)$ for some finite $\Lambda\}$.

   (b) If $\xi_\Lambda, \eta_\Lambda \in \Omega_\Lambda$, the function $\xi_{L\backslash\Lambda} \mapsto V(\xi_\Lambda \vee \xi_{L\backslash\Lambda}, \eta_\Lambda \vee \xi_{L\backslash\Lambda}$ is continuous on $\{\xi_{L\backslash\Lambda} \in \Omega_{L\backslash\Lambda} : \xi_\Lambda \vee \xi_{L\backslash\Lambda} \in \Omega$ and $\eta_\Lambda \vee \xi_{L\backslash\Lambda} \in \Omega\}$.

   (c) $V(\xi, \eta) + V(\eta, \zeta) + V(\zeta, \xi) = 0$ whenever the left-hand side is defined.

   Check that $(*)$ establishes a one-to-one correspondence between systems of conditional probabilities and objects V satisfying (a), (b), (c). These

* See the reduction to transitive and mixing systems in Theorems 5.2, 5.3. Another example is the use of "contours" to study the Ising model in zero magnetic field.

objects form a linear space. Check that the natural action of F^* on the V's is linear (F: a morphism).

4. Consider two lattice systems, with configuration spaces Ω and Ω'; let Φ, Φ' be interactions for them. Give natural definitions of sum and product of the two lattice systems: these are lattice systems with configuration spaces $\Omega^* = \Omega \cup \Omega'$ (disjoint union) and $\Omega^{**} = \Omega \times \Omega'$ (product) respectively. Define also sum and product of Φ and Φ'.

   (a) Show that the injection $\Omega \mapsto \Omega^*$ is a morphism.
   (b) Show that if σ, σ' are Gibbs states for Φ and Φ', then $\sigma \otimes \sigma'$ is a Gibbs state for $\Phi \times \Phi'$.

[To define the sum $(L^*, (\Omega^*_x)_{x\in L^*}, (\bar{\Omega}^*_\Lambda)_{\Lambda\in\mathcal{F}^*})$, use an arbitrary identification of the countable infinite sets L and L', and take $L^* = L = L'$, $\Omega^*_x = \Omega_x \cup \Omega'_x$. Choose $(\bar{\Omega}^*_\Lambda)_{\Lambda\in\mathcal{F}^*}$ so that $\xi \in \Omega^*$ only if $\xi \in \Omega$ or $\xi \in \Omega'$. To define the product, let $L^{**} = L \cup L'$ (disjoint union).]

3

Translation invariance. Theory of equilibrium states

In this chapter we assume invariance under translations, and we develop the theory of equilibrium states and of the pressure. General results on phase transitions are also obtained.

3.1 Translation invariance

The theory of Gibbs states has very interesting developments when invariance under a "sufficiently large" symmetry group G is assumed. We shall take $G = \mathbf{Z}^\nu$, $\nu \geqslant 1$, but remark that other groups may be of interest.* In the present chapter we shall not consider Gibbs states, but develop instead the theory of equilibrium states. The relation between Gibbs states and equilibrium states will be discussed in Chapter 4.

We take $L = \mathbf{Z}^\nu$ and let $G = \mathbf{Z}^\nu$ act on L by translations

$$(a, x) \mapsto a + x.$$

Furthermore we take $\Omega_x = \Omega_o$ for all $x \in \mathbf{Z}^\nu$, so that $\Pi_{x \in \mathbf{Z}^\nu} \Omega_x = (\Omega_o)^{\mathbf{Z}^\nu}$. For all $S \subset L$, define maps $\tau^a : \Pi_{x \in S} \Omega_x \mapsto \Pi_{x \in S-a} \Omega_x$ by

$$(\tau^a \xi)_x = \xi_{x+a}.$$

In this chapter we do not introduce the family $(\bar{\Omega}_\Lambda)_{\Lambda \in \mathfrak{F}}$ of Chapter 1, and simply assume that $\Omega \subset \Pi_{x \in \mathbf{Z}^\nu} \Omega_x$ and $\Omega_\Lambda \subset \Pi_{x \in \Lambda} \Omega_x$ (for Λ finite $\subset \mathbf{Z}^\nu$) are defined such that

(a) $\Omega_{\Lambda - a} = \tau^a \Omega_\Lambda$.
(b) $\Omega_\Lambda \supset \Omega_M | \Lambda$ if $\Lambda \subset M$.
(c) $\Omega \neq \emptyset$, and $\xi \in \Omega$ if and only if $\xi | \Lambda \subset \Omega_\Lambda$ for all Λ.

* Cf. the hierarchial model of Dyson [1].

As far as Ω is concerned these conditions amount to assuming that Ω *is an arbitrary non-empty closed τ-invariant subset of* $\Pi_{x\in\mathbf{Z}^\nu}\Omega_x$.

We let α_Λ, $\alpha_{\Lambda M}$, $\mathcal{C}$, $\mathcal{C}_\Lambda$ be as in Chapter 1. Notice that the τ^a are homeomorphisms of the compact set Ω. For each a, an automorphism of the algebra $\mathcal{C}$ is defined by $A \mapsto A\circ\tau^a$. This automorphism is an isometry and maps $\mathcal{C}_\Lambda$ onto $\mathcal{C}_{\Lambda+a}$.

An interaction Φ is (translation) invariant if it satisfies

$$\Phi(\tau^a\xi) = \Phi(\xi) \tag{3.1}$$

for all $a \in \mathbf{Z}^\nu$, $\xi \in \Omega_\Lambda$, Λ finite $\subset\mathbf{Z}^\nu$. These interactions form a Banach space $\mathcal{A}$ with respect to the norm

$$|\Phi| = \sum_{X\ni O}\frac{1}{|X|}\sup_{\xi\in\Omega_x}|\Phi(\xi)|. \tag{3.2}$$

We shall use the dense linear space $\mathcal{A}_o \subset \mathcal{A}$ consisting of *finite range interactions*; Φ is a finite range interaction if there exists Δ finite such that $\Phi(\xi|X) = 0$ unless $X - x \subset \Delta$ whenever $x \in X$.

In the present setup we may give $L = \mathbf{Z}^\nu$ the physical interpretation of a ν-dimensional crystal lattice. The assumed group invariance is that under lattice translations.

3.2 The function A_Φ

For each interaction $\Phi \in \mathcal{A}$ we now introduce a continuous function A_Φ on Ω. We want $A_\Phi \in \mathcal{C}$ to be so defined that $-A_\Phi(\xi)$ can be physically interpreted as the contribution of one lattice site (say 0) to the energy in the configuration ξ. This is achieved by writing

$$A_\Phi(\xi) = -\sum_{X\ni 0}\frac{1}{|X|}\Phi(\xi|X).$$

Other definitions are possible, however, with the same physical interpretation and giving the same value to $\sigma(A_\Phi)$ when σ is an invariant state (see Section 3.5 below). We could thus take

$$A_\Phi(\xi) = -\sum_X{}^*\,\Phi(\xi|X) \tag{3.3}$$

where $\sum^*$ extends over those X such that, if X is lexicographically ordered, 0 is its first element (or 0 is its last element). Lexicographic order is here any total order on $\mathbf{Z}^\nu$ which is compatible with the translations of $\mathbf{Z}^\nu$. We could also define A_Φ by (3.3) where $\sum^*$ extends over those X such that 0 is the

$[(|X|+1)/2]$th element* ("middle element") of X in lexicographic order. We shall use this last definition in what follows. It has the virtue that

$$\{A_\Phi : \Phi \in \mathcal{A}_o\} = \bigcup_{\Lambda \text{ finite}} \mathcal{C}_\Lambda \quad \text{and} \quad \{A_\Phi : \Phi \in \mathcal{A}\} = \mathcal{C}.$$

To see that each $A \in \mathcal{C}_\Lambda$ is of the form A_Φ, choose X such that $X \supset \Lambda$ and 0 is the $[(|X|+1)/2]$th element of X (in lexicographic order). Define $\Phi(\xi|X) = -A(\xi)$ and $\Phi(\xi|Y) = 0$ if Y is not a translate of X. Then $|\Phi| = \|A\|$ and $A = A_\Phi$. For general $A \in \mathcal{C}$, we can write $A = \sum_n A_n$ where $A_n \in \mathcal{C}_{\Lambda_n}$ and $\sum_n \|A_n\| < +\infty$. Therefore if Φ_n is chosen as above, we have $A = A_\Phi$ with $\Phi = \sum_n \Phi_n$.

Notice that the map $\Phi \to A_\Phi$ is linear, and that

$$\|A_\Phi\| \leqslant \sum_{X \ni 0} \frac{1}{|X|} \sup_{\xi \in \Omega_X} |\Phi(\xi)| = |\Phi|.$$

More precisely

$$\|A\| = \inf\{|\Phi| : A = A_\Phi\}. \tag{3.4}$$

Thus we have a continuous linear map $\Phi \mapsto A_\Phi$ from $\mathcal{A}$ onto $\mathcal{C}$.

3.3 Partition functions

Given $S \subset \mathbf{Z}^\nu$, we let

$$\Omega^*_S = \{\xi : (\exists \xi^* \in \Omega)\xi = \xi^*|S\}.$$

If Λ is finite and $\Phi \in \mathcal{A}$, we define the *partition functions*

$$Z^\Phi_\Lambda = \sum_{\xi \in \Omega_\Lambda} \exp\left[-U^\Phi_\Lambda(\xi)\right],$$

$$Z^{*\Phi}_\Lambda = \sum_{\xi \in \Omega^*_\Lambda} \exp\left[-U^\Phi_\Lambda(\xi)\right].$$

If $A \in \mathcal{C}$ we define also

$$Z^*_\Lambda(A) = \sum_{\xi \in \Omega^*_\Lambda} \exp \sum_{x \in \Lambda} A(\tau^x \xi^*),$$

where, for each $\xi \in \Omega^*_\Lambda$, an arbitrary choice of $\xi^* \in \Omega$ has been made so that $\xi^*|\Lambda = \xi$.

* $[(|X|+1)/2]$ is the integer part of $(|X|+1)/2$.

Let $P_\Lambda^\Phi = |\Lambda|^{-1} \log Z_\Lambda^\Phi$, then

$$\frac{d}{dt} P_\Lambda^{\Phi+t\Psi} = \left(Z_\Lambda^{\Phi+t\Psi}\right)^{-1} \sum_{\xi\in\Omega_\Lambda} \frac{-U_\Lambda^\Psi(\xi)}{|\Lambda|} \exp\left[-U_\Lambda^{\Phi+t\Psi}(\xi)\right],$$

and therefore

$$\begin{aligned} &|\Lambda| \frac{d^2}{dt^2} P(\Phi + t\Psi)\bigg|_{t=0} \\ &= \left(Z_\Lambda^\Phi\right)^{-2} \sum_{\xi\in\Omega_\Lambda} \sum_{\eta\in\Omega_\Lambda} \tfrac{1}{2}\left[U_\Lambda^\Psi(\xi) - U_\Lambda^\Psi(\eta)\right]^2 \exp\left[-U_\Lambda^\Phi(\xi) - U_\Lambda^\Phi(\eta)\right] \\ &\geqslant 0. \end{aligned} \tag{3.5}$$

This proves that $\Phi \mapsto P_\Lambda^\Phi$ is convex. On the other hand

$$\frac{d}{dt} P_\Lambda^{\Phi+t\Psi} \leqslant |\Psi|;$$

hence

$$\left|P_\Lambda^\Phi - P_\Lambda^\Psi\right| \leqslant \sup_{o\leqslant t\leqslant 1} \left|\frac{d}{dt} P^{\Phi+t(\Psi-\Phi)}\right| \leqslant |\Phi - \Psi|. \tag{3.6}$$

Notice also that

$$-|\Phi| \leqslant P_\Lambda^\Phi \leqslant |\Phi| + \log|\Omega_o|. \tag{3.7}$$

Properties similar to (3.5), (3.6), and (3.7) hold if Z_Λ^Φ is replaced by $Z_\Lambda^{*\Phi}$ or $Z_\Lambda^*(A)$. In particular let $P_\Lambda^*(A) = |\Lambda|^{-1} \log Z_\Lambda^*(A)$; then $A \mapsto P_\Lambda^*(A)$ is convex and

$$|P_\Lambda^*(A) - P_\Lambda^*(B)| \leqslant \|A - B\|, \tag{3.8}$$

where some fixed choice of the $\xi^* \in \Omega$ has been made in the definition of Z_Λ^*.

3.4 Theorem

Given $a_1, \ldots, a_\nu > 0$, define

$$\Lambda(a) = \{x \in \mathbf{Z}^\nu : 0 \leqslant x_i < a_i\}$$

and write $a \to \infty$ for $a_1, \ldots, a_\nu \to +\infty$. If $\Phi \in \mathfrak{A}$, $A \in \mathcal{C}$, the following limits exist:

$$P^\Phi = \lim_{a\to\infty} |\Lambda(a)|^{-1} \log Z_{\Lambda(a)}^\Phi = \lim_{a\to\infty} |\Lambda(a)|^{-1} \log Z_{\Lambda(a)}^{*\Phi}, \tag{3.9}$$

$$P(A) = \lim_{a\to\infty} |\Lambda(a)|^{-1} \log Z_{\Lambda(a)}^*(A). \tag{3.10}$$

Furthermore $P^\Phi = P(A_\Phi)$. *The function P, called the pressure, is convex and continuous on* $\mathcal{C}$; *in fact*

$$|P(A) - P(B)| \leqslant \|A - B\|. \tag{3.11}$$

If $t \in \mathbf{R}$ *we have also*

$$P(A + B \circ \tau^x - B + t) = P(A) + t. \tag{3.12}$$

We shall see later that the thermodynamic limit $a \to \infty$ in (9), (10) can be replaced by a more general limit $\Lambda \uparrow \infty$ (Section 3.9 and Corollary 3.13). For further properties of P see Section 6.8.

Let first $\Phi \in \mathcal{A}_o$: if $\xi \in \Omega_X$ and $X - X \not\subset \Delta$, then $\Phi(\xi) = 0$. If $\Lambda \subset M$ we have

$$Z_M \leqslant Z_\Lambda (|\Omega_o| e^{\|\Phi\|})^{|M| - |\Lambda|}, \tag{3.13}$$

where

$$\|\Phi\| = \sum_{X \ni 0} \sup_{\xi \in \Omega_x} |\Phi(\xi)|.$$

If $\Lambda_1 \cap \Lambda_2 = \emptyset$ we have also

$$Z_{\Lambda_1 \cup \Lambda_2} \leqslant Z_{\Lambda_1} Z_{\Lambda_2} e^{N(\Lambda_2)\|\Phi\|}, \tag{3.14}$$

where $N(\Lambda_2)$ is the number of points $x \in \Lambda_2$ such that $x + \Delta \not\subset \Lambda_2$.

Let

$$P^\Phi = \liminf_{a \to \infty} |\Lambda(a)|^{-1} \log Z^\Phi_{\Lambda(a)}. \tag{3.15}$$

Given $\varepsilon > 0$, we can choose b such that

$$\frac{N(\Lambda(b))}{|\Lambda(b)|} \|\Phi\| < \frac{\varepsilon}{2},$$

$$|\Lambda(b)|^{-1} \log Z^\Phi_{\Lambda(b)} < P^\Phi + \frac{\varepsilon}{2}.$$

Therefore if $a_1, \ldots, a_\nu$ are integral multiples of $b_1, \ldots, b_\nu$, use of (3.14) yields

$$|\Lambda(a)|^{-1} \log Z^\Phi_{\Lambda(a)} < P^\Phi + \varepsilon.$$

It is then easy to conclude, using (3.13), that

$$\limsup_{a \to \infty} |\Lambda(a)|^{-1} \log Z^\Phi_{\Lambda(a)} \leqslant P^\Phi. \tag{3.16}$$

Therefore, by (3.15),

$$\lim_{a \to \infty} |\Lambda(a)|^{-1} \log Z^\Phi_{\Lambda(a)} = P^\Phi. \tag{3.17}$$

The case of general $\Phi \in \mathcal{A}$ follows from the equicontinuity property (3.6).

Notice now that (3.14) may be replaced by

$$Z_{\Lambda_1 \cup \Lambda_2} \leqslant \tilde{Z}_{\Lambda_1} \tilde{Z}_{\Lambda_2} e^{N(\Lambda_2)\|\Phi\|}, \tag{3.18}$$

where

$$\tilde{Z}_{\Lambda_1} = \sum_{\xi \in \tilde{\Omega}_{\Lambda_1}} \exp[-U_\Lambda(\xi)],$$

$$\tilde{\Omega}_{\Lambda_1} = \{\tilde{\xi} | \Lambda_1 : \tilde{\xi} \in \Omega_{\Lambda_1 \cup \Lambda_2}\};$$

and similarly for $\tilde{Z}_{\Lambda_2}$. A generalization of (3.18) where $\Lambda_1 \cup \Lambda_2$ is replaced by $\Lambda_1 \cup \cdots \cup \Lambda_n$ is easily obtained. Taking $\Lambda_1, \ldots, \Lambda_n$ to be translates of $\Lambda(b)$ and $\Lambda_1 \cup \cdots \cup \Lambda_n = \Lambda(a)$, we find

$$\lim_{a \to \infty} |\Lambda(a)|^{-1} \log Z_{\Lambda(a)} \leqslant |\Lambda(b)|^{-1} \log \tilde{Z}_{\Lambda(b)} + \frac{N(\Lambda(b))}{|\Lambda(b)|} \|\Phi\|, \tag{3.19}$$

where $\tilde{Z}_{\Lambda(b)}$ is now computed with

$$\tilde{\Omega}_{\Lambda(b)} = \{\tilde{\xi} | \Lambda(b) : \tilde{\xi} \in \Omega_{\Lambda(b)+\Delta}\},$$

and Δ is any finite set containing 0. [To see this notice that $\Lambda_j + \Delta \subset \Lambda(a)$ for most j between 1 and n, when $a \to \infty$.] Since (3.19) holds for all Δ, we also have

$$P^\Phi \leqslant |\Lambda(b)|^{-1} \log Z^{*\Phi}_{\Lambda(b)} + \frac{N(\Lambda(b))}{|\Lambda(b)|} \|\Phi\|;$$

and since $Z^*_{\Lambda(b)} \leqslant Z_{\Lambda(b)}$, we obtain

$$\lim_{a \to \infty} |\Lambda(a)|^{-1} \log Z^{*\Phi}_{\Lambda(a)} = P^\Phi,$$

concluding the proof of (3.9).

If $\xi^* \in \Omega$, (3.3) gives

$$U^\Phi_\Lambda(\xi^* | \Lambda) + \sum_{x \in \Lambda} A_\Phi(\tau^x \xi^*) = \sum_{X \subset \Lambda} \Phi(\xi^* | X) - \sum_{x \in \Lambda} \sum_X{}^* \Phi(\xi^* | X + x).$$

Therefore, if $\Phi \in \mathcal{Q}_0$,

$$\left| U^\Phi_\Lambda(\xi^* | \Lambda) + \sum_{x \in \Lambda} A_\Phi(\tau^x \xi^*) \right| \leqslant N(\Lambda) |\Phi|$$

and

$$\lim_{a \to \infty} |\Lambda(a)|^{-1} \left[\log Z^{*\Phi}_{\Lambda(a)} - \log Z^*_{\Lambda(a)}(A_\Phi) \right] = 0.$$

By the equicontinuity property (3.8) and the density of the A_Φ, $\Phi \in \mathcal{Q}_0$, in $\mathcal{C}$ we obtain (3.10) and $P^\Phi = P(A_\Phi)$ for all $\Phi \in \mathcal{Q}$. The convexity of P and

(3.11) follow from the corresponding properties of P^*_Λ. One checks readily the properties

$$\lim_{a\to\infty} [P^*_{\Lambda(a)}(A_\Phi + A_\Psi \circ \tau^x - A_\Psi) - P^*_{\Lambda(a)}(A_\Phi)] = 0, \qquad \Phi, \Psi \in \mathcal{C}_0,$$

$$Z^*_{\Lambda(a)}(A + t) = e^{t|\Lambda|} Z^*_{\Lambda(a)}(A),$$

from which (3.12) follows.

3.5 Invariant states

For each $a \in \mathbf{Z}^\nu$ a linear map τ^a in the space $\mathcal{C}^*$ of real measures on Ω is defined by

$$(\tau^a\sigma)(A) = \sigma(A \circ \tau^a), \qquad \sigma \in \mathcal{C}^*, \quad A \in \mathcal{C}.$$

$\tau^a : \mathcal{C}^* \mapsto \mathcal{C}^*$ is continuous for the topology of $\mathcal{C}^*$ as weak dual of $\mathcal{C}$ (vague topology) and maps the set E of probability measures (states) onto itself. We let I be the set of τ-invariant states:

$$I = \{\sigma \in E : \tau^a\sigma = \sigma \ \text{ for all } a \in \mathbf{Z}^\nu\}.$$

We also say that these states are translation invariant states, or simply invariant states.

3.6 Proposition

The set I of invariant states is convex and compact; it is a Choquet simplex.

This is a general property of the set of probability measures invariant under a group of homeomorphisms of a compact set. It is readily seen that I is convex and compact. If σ is a τ-invariant measure, then $|\sigma|$ is also τ-invariant. From this it follows that I is a simplex (see Appendix A.5.5).

The extremal points of I are called *ergodic states*. The unique decomposition of an invariant state ρ into ergodic states is called the *ergodic decomposition*; it is given by a probability measure m_ρ on I such that

$$m_\rho(\hat{A}^2) = \lim_{\Lambda\nearrow\infty} \rho\left[\left(|\Lambda|^{-1}\sum_{x\in\Lambda} A\circ\tau^x\right)^2\right],$$

where $\hat{A}$ is the function on I defined by $\hat{A}(\sigma) = \sigma(A)$ (see Appendix A.5.6). An invariant state ρ is thus ergodic if and only if, for all $A \in \mathcal{C}$,

$$\lim_{\Lambda \nearrow \infty} \rho \left[\left(|\Lambda|^{-1} \sum_{x \in \Lambda} A \circ \tau^x \right)^2 \right] = [\rho(A)]^2.$$

This is a *weak cluster property.* Its physical interpretation is discussed in Section 3.15 below.

3.7 Theorem

Given $A \in \mathcal{C}$, let $I_A \subset \mathcal{C}^*$ be the set of measures σ on Ω such that

$$P(A + B) \geqslant P(A) + \sigma(B) \quad \textit{for all} \quad B \in \mathcal{C}.$$

(a) *$\emptyset \neq I_A \subset I$; I_A is convex and compact. We shall see later that I_A is a Choquet simplex and a face of I (Corollary* 3.14*).*

(b) *The set*

$$D = \{A \in \mathcal{C} : I_A \text{ consists of a single point}\}$$

is residual in $\mathcal{C}$.

Let now $\mathcal{X}$ be a separable Banach space and $\varphi : \mathcal{X} \mapsto \mathcal{C}$ a continuous linear map such that $\varphi\mathcal{X}$ is dense in $\mathcal{C}$.

(c) *For $\Phi \in \mathcal{X}$, define*

$$I'_\Phi = \{F \in \mathcal{X}^* : P \circ \varphi(\Phi + \Psi) \geqslant P \circ \varphi(\Phi) + F(\Psi) \textit{ for all } \Psi \in \mathcal{X}\},$$
$$D' = \{\Phi \in \mathcal{X} : I'_\Phi \text{ consists of a single point}\}.$$

Then $I'_\Phi = \{\sigma \circ \varphi : \sigma \in I_{\varphi\Phi}\}$, and $D' = \varphi^{-1} D$ is residual in $\mathcal{X}$.

(d) *$I_{\varphi\Phi}$ is the closed convex hull of the set of ρ such that*

$$\rho = \lim_{n \to \infty} \rho_n, \qquad \rho_n \in I_{\varphi\Phi_n},$$
$$\lim_{n \to \infty} \|\Phi_n - \Phi\| = 0, \qquad \Phi_n \in \varphi^{-1} D.$$

The elements of I_A are called *equilibrium states for A*; the elements of I_{A_Φ} are the *equilibrium states for the interaction Φ*.

$I_A \neq \emptyset$ and (b) hold for any convex continuous function P on a separable Banach space (see Appendix A.3.6 and A.3.7). Therefore also the set D' is residual in $\mathcal{X}$.

If $F \in I'_\Phi$ we have, using (3.11),

$$F(\Psi) \leqslant P \circ \varphi(\Phi + \Psi) - P \circ \varphi(\Phi) \leqslant \|\varphi\Psi\|;$$

hence, replacing Ψ by $-\Psi$, we obtain $|F(\Psi)| \leqslant \|\varphi\Psi\|$, or

$$\sup_{\|\varphi\Psi\|\leqslant 1} |F(\Psi)| \leqslant 1; \tag{3.20}$$

and there exists $\sigma \in \mathcal{C}^*$ such that $F = \sigma \circ \varphi$. By the density of $\varphi\mathfrak{X}$ in $\mathcal{C}$, σ is unique, and $\sigma \in I_{\varphi\Phi}$. Thus $I'_\Phi \subset \{\sigma \circ \varphi : \sigma \in I_{\varphi\Phi}\}$ and therefore $I'_\Phi = \{\sigma \circ \varphi : \sigma \in I_{\varphi\Phi}\}$. In particular I'_Φ consists of a single point if and only if $I_{\varphi\Phi}$ does (use the density of $\varphi\mathfrak{X}$ in $\mathcal{C}$) : $D' = \varphi^{-1} D$. Since we know that D' is residual, (c) is proved.

To prove (a), we already know that $I_A \neq \emptyset$ and, taking $\mathfrak{X} = \mathcal{C}$ in (3.20), that $\|\sigma\| \leqslant 1$ if $\sigma \in I_A$. We also have, by (3.12),

$$\sigma(1) = -\sigma(-1) \geqslant -[P(A-1) - P(A)] = 1.$$

But $\|\sigma\| \leqslant 1$ and $\sigma(1) \geqslant 1$ imply $\sigma \geqslant 0$ and $\|\sigma\| = 1$, i.e., $\sigma \in E$. Furthermore, by (3.12) again,

$$\begin{aligned} 0 = P(A + B \circ \tau^x - B) - P(A) &\geqslant \sigma(B \circ \tau^x - B) \\ &\geqslant -[P(A - B \circ \tau^x + B) - P(A)] = 0, \end{aligned}$$

so that $\sigma \in I$. Thus $I_A \subset I$. Clearly I_A is convex and compact, concluding the proof of (a).

Let ρ, ρ_n, Φ, Φ_n be as in (d). We have

$$P(\varphi\Phi_n + B) \geqslant P(\varphi\Phi_n) + \rho_n(B)$$

for all $B \in \mathcal{C}$; hence

$$P(\varphi\Phi + B) \geqslant P(\varphi\Phi) + \rho(B),$$

showing that $\rho \in I_{\varphi\Phi}$.

Suppose now that $\sigma \in I_{\varphi\Phi}$ were not in the closed convex hull of those ρ. By the separation theorem for compact sets (see Appendix A.3.3(c)) there would then exist $\Psi \in \mathfrak{X}$ such that

$$\sup_{\sigma \in I_{\varphi\Phi}} \sigma(\varphi\Psi) > \sup_{\rho} \rho(\varphi\Psi).$$

Let $\Phi_n = \Phi + (1/n)\Psi + X_n$, where $\|X_n\| < 1/n^2$ and $\Phi_n \in \varphi^{-1}D$. Write $I_{\varphi\Phi_n} = \{\rho_n\}$. By the convexity of P we have, if $\sigma \in I_{\varphi\Phi}$,

$$\sigma\left(\varphi\left(\frac{1}{n}\Psi + X_n\right)\right) \leqslant \rho_n\left(\varphi\left(\frac{1}{n}\Psi + X_n\right)\right);$$

hence

$$\sigma(\varphi\Psi) - \frac{\|\varphi\|}{n} \leqslant \rho_n(\varphi\Psi) + \frac{\|\varphi\|}{n};$$

and if ρ is a limit point of ρ_n,

$$\sigma(\varphi\Psi) \leqslant \rho(\varphi\Psi),$$

contradicting our assumption on Ψ. This concludes the proof of (d), which could also be obtained from a general property of convex functions (see Appendix A.3.7).

3.8 Entropy

Given a probability measure σ_Λ on Ω_Λ (for finite $\Lambda \subset \mathbf{Z}^\nu$) the corresponding *entropy* is defined by

$$S(\sigma_\Lambda) = -\sum_{\xi \in \Omega_\Lambda} \sigma_\Lambda\{\xi\} \log \sigma_\Lambda\{\xi\}.$$

One checks readily that

$$0 \leqslant S(\sigma_\Lambda) \leqslant |\Lambda| \log |\Omega_0|.$$

When $\Lambda = \emptyset$, the entropy vanishes.

If σ'_Λ is another probability measure, and $0 < \alpha < 1$, the following inequalities hold:

$$\begin{aligned} \alpha S(\sigma_\Lambda) + (1-\alpha) S(\sigma'_\Lambda) &\leqslant S(\alpha\sigma_\Lambda + (1-\alpha)\sigma'_\Lambda) \\ &\leqslant \alpha S(\sigma_\Lambda) + (1-\alpha) S(\sigma'_\Lambda) + \log 2. \end{aligned} \tag{3.21}$$

Writing $\sigma_\xi = \sigma_\Lambda\{\xi\}$, $\sigma'_\xi = \sigma'_\Lambda\{\xi\}$, we have indeed, using the convexity of $t \mapsto t \log t$ and the monotonicity of the logarithm,

$$\begin{aligned} &-\sum_\xi [\alpha\sigma_\xi \log \sigma_\xi + (1-\alpha)\sigma'_\xi \log \sigma'_\xi] \\ &\leqslant -\sum_\xi [\alpha\sigma_\xi + (1-\alpha)\sigma'_\xi] \log[\alpha\sigma_\xi + (1-\alpha)\sigma'_\xi] \\ &\leqslant -\sum_\xi [\alpha\sigma_\xi \log \alpha\sigma_\xi + (1-\alpha)\sigma'_\xi \log(1-\alpha)\sigma'_\xi] \\ &= -\sum_\xi [\alpha\sigma_\xi \log \sigma_\xi + (1-\alpha)\sigma'_\xi \log \sigma'_\xi] - \alpha \log \alpha - (1-\alpha)\log(1-\alpha) \\ &\leqslant -\sum_\xi [\alpha\sigma_\xi \log \sigma_\xi + (1-\alpha)\sigma'_\xi \log \sigma'_\xi] + \log 2. \end{aligned}$$

Let now $\sigma \in E$, then $S(\alpha_\Lambda \sigma)$ is an *increasing* function of Λ, and satisfies the *strong subadditivity* property

$$S(\alpha_{\Lambda_1 \cup \Lambda_2}\sigma) + S(\alpha_{\Lambda_1 \cap \Lambda_2}\sigma) \leqslant S(\alpha_{\Lambda_1}\sigma) + S(\alpha_{\Lambda_2}\sigma). \tag{3.22}$$

The increase follows directly from the monotonicity of the logarithm. To prove (3.22) we shall use the inequality $-\log(1/t) \leqslant t - 1$. We have

$$\begin{aligned}
&S(\alpha_{\Lambda_1\cup\Lambda_2}\sigma) + S(\alpha_{\Lambda_1\cap\Lambda_2}\sigma) - S(\alpha_{\Lambda_1}\sigma) - S(\alpha_{\Lambda_2}\sigma) \\
&\quad = -\sum_{\xi\in\Omega_{\Lambda_1\cap\Lambda_2}}\ \sum_{\xi'\in\Omega_{\Lambda_1/\Lambda_2}}\ \sum_{\xi''\in\Omega_{\Lambda_2/\Lambda_1}} \sigma_{\xi\vee\xi'\vee\xi''} \log \frac{\sigma_{\xi\vee\xi'\vee\xi''}\sigma_\xi}{\sigma_{\xi\vee\xi'}\sigma_{\xi\vee\xi''}} \\
&\quad \leqslant \sum_{\xi\xi'\xi''} \sigma_{\xi\vee\xi'\vee\xi''} \left[\frac{\sigma_{\xi\vee\xi'}\sigma_{\xi\vee\xi''}}{\sigma_{\xi\vee\xi'\vee\xi''}\sigma_\xi} - 1\right] \\
&\quad = \sum_{\xi\xi'} \frac{\sigma_{\xi\vee\xi'}}{\sigma_\xi} \sum_{\xi''} \sigma_{\xi\vee\xi''} - \sum_{\xi\xi'\xi''} \sigma_{\xi\vee\xi'\vee\xi''} = \sum_{\xi\xi'} \sigma_{\xi\vee\xi'} - 1 = 0.
\end{aligned}$$

(To avoid undefined expressions one may, in the above calculation, first assume $\sigma_{\xi\vee\xi'\vee\xi''} > 0$ for all ξ, ξ', ξ'', and then take a limit.)

3.9 Infinite limit in the sense of van Hove

We say that the finite sets $\Lambda \subset \mathbf{Z}^\nu$ *tend to infinity in the sense of van Hove* (and we write $\Lambda \nearrow \infty$) if $|\Lambda| \to \infty$ and, for each $a \in \mathbf{Z}^\nu$,

$$\frac{|(\Lambda + a)/\Lambda|}{|\Lambda|} \to 0.$$

Roughly speaking this means that the "boundary of Λ" becomes negligible in the limit as compared to Λ.

3.10 Theorem

If $\sigma \in I$, the following limit exists:

$$s(\sigma) = \lim_{\Lambda\nearrow\infty} |\Lambda|^{-1} S(\alpha_\Lambda\sigma) = \inf_\Lambda |\Lambda|^{-1} S(\alpha_\Lambda\sigma).$$

The function s, called the (mean) entropy, is $\geqslant 0$, affine, and upper semicontinuous on I.

If we take Λ_1 and Λ_2 disjoint, (3.22) becomes subadditivity:

$$S(\alpha_{\Lambda_1\cup\Lambda_2}\sigma) \leqslant S(\alpha_{\Lambda_1}\sigma) + S(\alpha_{\Lambda_2}\sigma). \tag{3.23}$$

Assuming $\sigma \in I$, we have also

$$S(\alpha_{(\Lambda+x)}\sigma) = S(\alpha_\Lambda\sigma). \tag{3.24}$$

Defining $\Lambda(a)$ as in Theorem 3.4, we let

$$s = \inf_a |\Lambda(a)|^{-1} S(\alpha_{\Lambda(a)}\sigma) \tag{3.25}$$

and, given $\varepsilon > 0$, choose b such that

$$|\Lambda(b)|^{-1} S(\alpha_{\Lambda(b)}\sigma) \leqslant s + \varepsilon.$$

If we translate $\Lambda(b)$ by elements of

$$\mathbf{Z}^\nu(b) = \{x \in \mathbf{Z}^\nu : x_i = n_i b_i, n_i \in \mathbf{Z}^\nu, i = 1, \ldots, \nu\},$$

we obtain a partition of $\mathbf{Z}^\nu$. Let Λ_+ be the union of those translates which have non-empty intersection with Λ. Then $\Lambda_+ \supset \Lambda$ and $|\Lambda_+|/|\Lambda| \to 1$ when $\Lambda \nearrow \infty$. Since $S(\alpha_\Lambda\sigma)$ is an increasing function of Λ, and (3.23), (3.24) hold,

$$\mathrm{S}(\alpha_\Lambda\sigma) \leqslant S(\alpha_{\Lambda_+}\sigma) \leqslant \frac{|\Lambda_+|}{|\Lambda(b)|} S(\alpha_{\Lambda(b)}\sigma) \leqslant |\Lambda_+|(s + \varepsilon);$$

therefore

$$\limsup_{\Lambda\nearrow\infty} |\Lambda|^{-1} S(\alpha_\Lambda\sigma) \leqslant s + \varepsilon. \tag{3.26}$$

From (3.25) and (3.26) we obtain in particular

$$\lim_{a\to\infty} |\Lambda(a)|^{-1} S(\alpha_{\Lambda(a)}\sigma) = s.$$

Strong subadditivity (3.22) shows that

$$S(\alpha_{\Lambda\cup\{x\}}\sigma) - S(\alpha_\Lambda\sigma) \geqslant S(\alpha_{\Lambda'\cup\{x\}}\sigma) - S(\alpha_{\Lambda'}\sigma) \tag{3.27}$$

when $x \notin \Lambda' \supset \Lambda$. This permits an estimate of the increase in the entropy for a set Λ to which points are added successively in lexicographic order. In particular, if Λ is fixed (up to translations) and one takes for Λ' the sets successively obtained in the lexicographic construction of a large $\Lambda(a)$, (3.27) holds for most Λ'. Therefore

$$S(\alpha_{\Lambda\cup\{x\}}\sigma) - S(\alpha_\Lambda\sigma) \geqslant \lim_{a\to\infty} |\Lambda(a)|^{-1} S(\alpha_{\Lambda(a)}\sigma) = s.$$

Hence

$$S(\alpha_\Lambda\sigma) \geqslant |\Lambda| s$$

for all Λ. From this and (3.26) we get

$$\lim_{\Lambda\nearrow\infty} |\Lambda|^{-1} S(\alpha_\Lambda\sigma) = \inf_\Lambda |\Lambda|^{-1} S(\alpha_\Lambda\sigma) = s,$$

proving the first part of the theorem.

Since $S \geqslant 0$, we also have $s \geqslant 0$. That s is affine follows from (3.21). Finally s is upper semicontinuous as the inf of the continuous functions $\sigma \mapsto |\Lambda|^{-1} S(\alpha_\Lambda \sigma)$.

3.11 Lemma

If E_Λ is the set of all probability measures on Ω_Λ, then

$$\log Z_\Lambda^\Phi = \max_{\sigma_\Lambda \in E_\Lambda} \left[S(\sigma_\Lambda) - \sigma_\Lambda\big(U_\Lambda^\Phi\big) \right].$$

We have indeed, by the concavity of the logarithm,

$$S(\sigma_\Lambda) - \sigma_\Lambda\big(U_\Lambda^\Phi\big) = \sum_{\xi \in \Omega_\Lambda} \sigma_\Lambda\{\xi\} \log \frac{e^{-U_\Lambda(\xi)}}{\sigma_\Lambda\{\xi\}} \leqslant \log \sum_{\xi \in \Omega_\Lambda} e^{-U_\Lambda(\xi)};$$

and equality holds if $\sigma_\Lambda\{\xi\} = (Z_\Lambda^\Phi)^{-1} e^{-U_\Lambda(\xi)} = \mu_{(\Lambda)}\{\xi\}$.

The following theorem gives a version of this *variational principle* in the thermodynamic limit $\Lambda \nearrow \infty$.

3.12 Theorem

For all $A \in \mathcal{C}$,

$$P(A) = \max_{\sigma \in I} [s(\sigma) + \sigma(A)]; \tag{3.28}$$

and the maximum is reached precisely on I_A. For all $\sigma \in I$,

$$s(\sigma) = \inf_{A \in \mathcal{C}} [P(A) - \sigma(A)]. \tag{3.29}$$

We first prove that

$$P(A) = \sup_{\sigma \in I} [s(\sigma) + \sigma(A)] \tag{3.30}$$

when $A = A_\Phi$, $\Phi \in \mathcal{C}_0$. By Lemma 3.11 we have

$$P(A_\Phi) \geqslant s(\sigma) + \sigma(A_\Phi) \qquad \text{for all} \qquad \sigma \in I, \tag{3.31}$$

where we have used

$$\sigma(A_\Phi) = -\lim_{a \to \infty} |\Lambda(a)|^{-1} (\alpha_{\Lambda(a)} \sigma)\big(U_{\Lambda(a)}^\Phi\big). \tag{3.32}$$

Using the Gibbs ensembles $\mu_{(\Lambda)}$ for the interaction Φ, we now define

$$\rho_{\Lambda,n}\{\xi\} = |\Lambda(a_n)|^{-1} \sum_{x : \Lambda + x \subset \Lambda(a_n)} (\alpha_{\Lambda+x} \mu_{(\Lambda(a_n))})\{\tau^{-x} \xi\}.$$

It is readily seen that the sequence (a_n) can be chosen such that $a_n \to \infty$ and the following limit exists for each finite $\Lambda \subset \mathbf{Z}^\nu$:

$$\rho_\Lambda = \lim_{n\to\infty} \rho_{\Lambda,n}. \tag{3.33}$$

The unique state ρ such that $\rho_\Lambda = \alpha_\Lambda \rho$ for all Λ then belongs to I, and we have

$$\begin{aligned} s(\rho) &= \lim_{b\to\infty} |\Lambda(b)|^{-1} S(\rho_{\Lambda(b)}) \\ &= -\lim_{b\to\infty}\lim_{n\to\infty} |\Lambda(b)|^{-1} \sum_{\xi\in\Omega_{\Lambda(b)}} \rho_{\Lambda(b),n}\{\xi\} \log \rho_{\Lambda(b),n}\{\xi\} \\ &\geqslant \limsup_{b\to\infty}\limsup_{n\to\infty} |\Lambda(b)|^{-1}|\Lambda(a_n)|^{-1} \sum_{x:\Lambda(b)+x\subset\Lambda(a_n)} S(\alpha_{\Lambda(b)+x}\mu_{(\Lambda(a_n))}) \\ &\geqslant \limsup_{n\to\infty} |\Lambda(a_n)|^{-1} S(\mu_{(\Lambda(a_n))}). \end{aligned} \tag{3.34}$$

[In the last step we used (3.23) and the fact that $\Lambda(a_n)$ can be written in $|\Lambda(b)|$ ways as a union of translates of $\Lambda(b)$ plus k_n points where $k_n/|\Lambda(a_n)| \to 0$ when $n \to \infty$.] From (3.34) we obtain

$$\begin{aligned} s(\rho) + \rho(A_\Phi) &\geqslant \lim_{n\to\infty} |\Lambda(a_n)|^{-1} \sum_{\xi\in\Omega_{\Lambda(a_n)}} \mu_{(\Lambda(a_n))}\{\xi\} \log \frac{\exp\left[-U^\Phi_{\Lambda(a_n)}(\xi)\right]}{\mu_{(\Lambda(a_n))}\{\xi\}} \\ &= \lim_{n\to\infty} |\Lambda(a_n)|^{-1} \log Z^\Phi_{\Lambda(a_n)} = P(A_\Phi). \end{aligned} \tag{3.35}$$

Equations (3.31) and (3.35) prove (3.30) when $A = A_\Phi$, $\Phi \in \mathcal{C}_0$. But since both sides of (3.30) are continuous in A, this relation holds, by density, for all $A \in \mathcal{C}$. Furthermore, since σ is upper semicontinuous, the sup is reached and (3.28) holds.

We want now to prove (3.29). We already know by (3.28) that

$$s(\sigma) \leqslant P(A) - \sigma(A),$$

and it remains to show that by proper choice of A the right-hand side becomes as close as desired to $s(\sigma)$. Let

$$C = \{(\sigma, t) \in \mathcal{C}^* \times \mathbf{R} : \sigma \in I \text{ and } 0 \leqslant t \leqslant s(\sigma)\}.$$

Since s is affine upper semi-continuous, C is convex and compact. Given $\rho \in I$ and $u > s(\rho)$, there exist (see Appendix A.3.3) $A \in \mathcal{C}$ and $c \in \mathbf{R}$ such that

$$-\rho(A) + c = u$$

and

$$-\sigma(A) + c > s(\sigma) \qquad \text{for all} \quad \sigma \in I;$$

hence

$$-\sigma(A) + u + \rho(A) > s(\sigma).$$

If σ is chosen such that $P(A) = s(\sigma) + \sigma(A)$, we obtain

$$\begin{aligned} 0 &\leqslant P(A) - s(\rho) - \rho(A) \\ &= s(\sigma) + \sigma(A) - s(\rho) - \rho(A) \\ &< u - s(\rho). \end{aligned}$$

The right-hand side is arbitrarily small, and (3.29) follows.

The condition $\rho \in I_A$, i.e.

$$P(A + B) \geqslant P(A) + \rho(B) \qquad \text{for all} \quad B \in \mathcal{C}$$

is equivalent to

$$P(A + B) - \rho(A + B) \geqslant P(A) - \rho(A) \qquad \text{for all} \quad B \in \mathcal{C},$$

or to

$$\inf_{C \in \mathcal{C}} [P(C) - \rho(C)] \geqslant P(A) - \rho(A),$$

or by (3.29) to

$$s(\rho) \geqslant P(A) - \rho(A);$$

hence the max of (3.28) is reached precisely on I_A, concluding the proof of the theorem.

3.13 Corollary

The formulas

$$P^\Phi = \lim_{\Lambda \nearrow \infty} |\Lambda|^{-1} \log Z^\Phi_\Lambda = \lim_{\Lambda \nearrow \infty} |\Lambda|^{-1} \log Z^{*\Phi}_\Lambda, \tag{3.36}$$

$$P(A) = \lim_{\Lambda \nearrow \infty} |\Lambda|^{-1} \log Z^*_\Lambda(A) \tag{3.37}$$

hold, generalizing (3.9) *and* (3.10).

We notice first that, if $\Phi \in \mathcal{A}_0$, the argument leading to (3.16) yields

$$\limsup_{\Lambda \nearrow \infty} |\Lambda|^{-1} \log Z^\Phi_\Lambda \leqslant P^\Phi.$$

On the other hand Lemma 3.11 shows that, for all $\sigma \in I$,

$$\liminf_{\Lambda \nearrow \infty} |\Lambda|^{-1} \log Z^\Phi_\Lambda \geqslant s(\sigma) + \sigma(A_\Phi),$$

so that

$$\lim_{\Lambda \nearrow \infty} |\Lambda|^{-1} \log Z_{\Lambda}^{\Phi} = P^{\Phi}$$

when $\Phi \in \mathcal{A}_0$. The general case $\Phi \in \mathcal{A}$ follows as usual by the density of $\mathcal{A}_0$ in $\mathcal{A}$ and equicontinuity. Replacing Ω_Λ by Ω_Λ^* we find that

$$\lim_{\Lambda \nearrow \infty} |\Lambda|^{-1} \log Z_{\Lambda}^{*\Phi}$$

exists. By Theorem 3.4 this limit is equal to P^{Φ}. This proves (3.36). The proof of (3.37) follows that of (3.10) in Theorem 3.4.

3.14 Corollary

For each $A \in \mathcal{C}$, I_A is a Choquet simplex, and a face of I.

We know that I is a simplex (Proposition 3.6). If $\rho \in I_A$, let m_ρ be the unique probability measure on I, carried by the extremal points of I, and with resultant ρ. Writing $\hat{A}(\sigma) = \sigma(A)$, we have (see Appendix A.5.1)

$$m_\rho(s + \hat{A}) = s(\rho) + \rho(A) = P(A).$$

Hence the support of m_ρ is contained in $\{\sigma \in I : s(\sigma) + \sigma(A) = P(A)\} = I_A$. This implies that I_A is a simplex, and a face of I.

3.15 Physical interpretation

We have indicated in Section 1.1 that an element A of $\mathcal{C}$ may be considered as an observable. If $\rho \in E$, the probability measure $\mu_A = A_\rho$ on $\mathbf{R}$ is defined by

$$\mu_A(\varphi) = \rho(\varphi \circ A) \quad \text{(all continuous } \varphi : \mathbf{R} \mapsto \mathbf{R}\text{)}.$$

μ_A describes the distribution of values of the observable A in the state ρ. In general A fluctuates, i.e., the support of μ_A consists of more than one point.

Consider now the average of A over translations in Λ defined by

$$\langle A \rangle_\Lambda = |\Lambda|^{-1} \sum_{x \in \Lambda} A \circ \tau^x.$$

Let $\rho \in I$. The condition

$$\lim_{\Lambda \nearrow \infty} \rho([\langle A \rangle_\Lambda - \rho(A)]^2) = 0 \quad \text{for all} \quad A \in \mathcal{C}$$

holds if and only if ρ is ergodic (see Section 3.6). This condition means that $\langle A\rangle_\Lambda$ fluctuates little for large Λ: we express this by saying that ρ is a (*pure thermodynamic*) *phase*. A phase is indeed characterized by the fact that "coarse-grained" quantities (i.e. the averages $\langle A\rangle_\Lambda$) do not fluctuate (in the limit $\Lambda \nearrow \infty$). On the other hand, for a *mixture* some coarse-grained quantities fluctuate. Notice that, on physical grounds, every mixture should have a unique decomposition into pure phases.

Let now ρ be an equilibrium state for A. Since I_A is a simplex, ρ has a unique decomposition into extremal points of I_A. Since I_A is a face of I, this decomposition is the same as the ergodic decomposition of ρ (see Section 3.6). This decomposition is physically interpreted as the decomposition of an equilibrium state into pure thermodynamic phases.

The residual set $D \subset \mathcal{C}$ (or $\varphi^{-1}D \subset \mathcal{X}$, where $\mathcal{X}$, φ are as in Theorem 3.7) may be considered as a "large" set. Therefore "in general" there exists only one pure thermodynamic phase associated with $A \in \mathcal{C}$ (or with $\Phi \in \mathcal{X}$). This is a weak form of the *Gibbs phase rule*.

3.16 Theorem

Given $A \in \mathcal{C}$ (respectively $\Phi \in \mathcal{A}$), $\sigma \in I$, and $\varepsilon > 0$, there exist $A' \in \mathcal{C}$ (respectively $\Phi' \in \mathcal{A}$) and $\sigma' \in I_{A'}$ such that

$$\|\sigma' - \sigma\| \leqslant \varepsilon$$

and

$$\|A' - A\| \leqslant \frac{1}{\varepsilon}[P(A) - \sigma(A) - s(\sigma)]$$

$$\left(\textit{respectively} \quad \|\Phi' - \Phi\| \leqslant \frac{1}{\varepsilon}[P^\Phi - \sigma(A_\Phi) - s(\sigma)]\right).$$

This is precisely the theorem of Bishop and Phelps given in Appendix A.3.6, with $V = \mathcal{C}$ (respectively $V = \mathcal{A}$: in this case notice that the tangent functionals are given by Theorem 3.7(c), and use (3.4) to get $\|\sigma' - \sigma\| \leqslant \varepsilon$).

3.17 Corollary

(a) *The union of the I_A, for $A \in \mathcal{C}$, i.e. the set of all equilibrium states, is dense in I for the norm topology.*

(b) *If $\rho_1, \ldots, \rho_n$ are ergodic states there is an interaction $\Phi \in \mathcal{A}$ for which these are all equilibrium states.*

(a) is clear. In particular, given $\rho_1, \ldots, \rho_n$ ergodic there is an equilibrium state ρ for some interaction $\Phi \in \mathcal{C}$ such that $\|\rho - (1/n)(\rho_1 + \ldots + \rho_n)\| < 1/n$. Let m_ρ give the ergodic decomposition of ρ; then we have

$$\left\| m_\rho - \frac{1}{n}(\delta_{\rho_1} + \cdots + \delta_{\rho_n}) \right\| < \frac{1}{n}$$

(see Appendix A.5.5), and therefore $m_\rho(\{\rho_1\}) > 0, \ldots, m_\rho(\{\rho_n\}) > 0$. Thus $\rho_1, \ldots, \rho_n$ are equilibrium states for Φ by Corollary 3.14.

3.18 Approximation of invariant states by equilibrium states

We have just seen that every invariant state σ can be approximated in norm by an equilibrium state σ' for some interaction in $\mathcal{C}$. This interesting result should however be viewed as a pathology from the physical point of view, because the interactions in $\mathcal{C}$ are not all physically acceptable. In fact, to be able to define Gibbs states we shall in the next chapter introduce a smaller space $\mathcal{B}$ of interactions. It will then be seen (Proposition 4.7(b)) that if $\Phi, \Phi' \in \mathcal{B}$ have an equilibrium state ρ in common, they are in some sense equivalent, and all their equilibrium states are the same. [This is quite different from the situation of Corollary 3.17(b).]

Physically meaningful results can be obtained by approximating invariant states by equilibrium states, using a general theorem on convex functions due to Israel (see Appendix A.3.6). With this theorem one shows that an interaction in a certain subspace or cone of $\mathcal{C}$ can be found, which has an equilibrium state satisfying certain inequalities. If these inequalities express that a certain cluster property does not hold, physical consequences can be derived. Theorem 3.20 below gives an example where it is shown that there are several distinct equilibrium states for the same interaction. (For another example see Exercise 1 of Chapter 4.)

3.19 Lemma

Let $A_1, A_2 \in \mathcal{C}$ and $S \subset \mathbf{Z}^\nu$. We define a convex cone

$$\mathcal{Q}_S = \left\{ a_1 A_1 + a_2 A_2 + \frac{1}{2} \sum_{x \in \mathbf{Z}^\nu} (b_x A_1 \cdot (A_2 \circ \tau^x) + b_{-x}(A_1 \circ \tau^x) \cdot A_2) \right.$$

$$\left. : a_1, a_2, b_x \in \mathbf{R}, b_x \geqslant 0, b_x = 0 \text{ if } x \notin S, \sum_{x \in S} b_x < +\infty \right\}.$$

We assume that $A_1, A_2 \in \mathcal{C}_\Delta$ *for some finite* $\Delta \subset \mathbf{Z}^\nu$. *Given* $\sigma_0 \in I$, $B_0 \in \mathcal{C}$, *and* $\varepsilon > 0$, *there exist* $B \in B_0 + \mathcal{Q}_S$ *and* $\sigma \in I_B$ *such that*

$$\|B - B_0\| \leqslant \frac{1}{\varepsilon}[P(B_0) - \sigma(B_0) - s(\sigma_0)] \tag{3.38}$$

and

$$\begin{aligned} &\sigma(A_1 \cdot (A_2 \circ \tau^x)) - \sigma(A_1)\sigma(A_2) \\ &\geqslant \sigma_0(A_1 \cdot (A_2 \circ \tau^x)) - \sigma_0(A_1)\sigma_0(A_2) - 3\varepsilon\|A_1\|\|A_2\| \end{aligned} \tag{3.39}$$

for all $x \in S$.

We may assume that 0 is the "middle" element of Δ in lexicographic order. Given $a_1, a_2, b_x \in \mathbf{R}$ such that $b_x \geqslant 0$, $b_x = 0$ if $x \in S$, and $\sum b_x < +\infty$, let $\Phi \in \mathcal{A}$ be defined by

$$\begin{aligned} -\Phi(\xi|\Delta) &= a_1 A_1(\xi) + a_2 A_2(\xi) + b_0 A_1(\xi) \cdot A_2(\xi), \\ -\Phi(\xi|\Delta \cup (\Delta + x)) & \\ &= b_x A_1(\xi) \cdot A_2(\tau^x \xi) + b_{-x} A_1(\tau^x \xi) \cdot A_2(\xi) \qquad \text{if} \quad x \neq 0, \\ \Phi(\xi|X) = 0 \qquad &\text{if} \quad X \text{ is not a translated of } \Delta \text{ or } \Delta \cup (\Delta + x). \end{aligned}$$

Using the expression (3.2) for the norm, we see that the set of these Φ is a closed convex cone $\mathcal{R} \subset \mathcal{A}$.

Let $\Psi_0 \in \mathcal{A}$ be such that $A_{\Psi_0} = B_0$. We now use Israel's theorem (Appendix A.3.6) and the characterization of tangent planes to $\Phi \to P^\Phi$ given in Theorem 3.7(c). We find thus $\Psi \in \Psi_0 + \mathcal{R}$, and $\sigma \in I_{A_\Psi}$, such that

$$|\Psi - \Psi_0| \leqslant \frac{1}{\varepsilon}[P^{\Psi_0} - \sigma_0(A_{\Psi_0}) - s(\sigma_0)] \tag{3.40}$$

and, for all $\Phi \in \mathcal{R}$,

$$\sigma(A_\Phi) \geqslant \sigma_0(A_\Phi) - \varepsilon|\Phi|. \tag{3.41}$$

From (3.41) one gets in particular

$$\sigma(A_1) - \sigma_0(A_1)| \leqslant \varepsilon\|A_1\|, \quad |\sigma(A_2) - \sigma_0(A_2)| \leqslant \varepsilon\|A_2\|, \tag{3.42}$$

and (for all $x \in S$)

$$\sigma(A_1 \cdot (A_2 \circ \tau^x)) \geqslant \sigma_0(A_1 \cdot (A_2 \circ \tau^x)) - \varepsilon\|A_1\| \cdot \|A_2\|. \tag{3.43}$$

(3.42) implies

$$|\sigma(A_1)\sigma(A_2) - \sigma_0(A_1)\sigma_0(A_2)| \leqslant 2\varepsilon\|A_1\| \cdot \|A_2\|$$

which, together with (3.43), proves (3.39).

Let C be the element of $\mathfrak{Q}_S$ formed with the same a_1, a_2, b_x which define $\Psi - \Psi_0$ in $\mathfrak{R}$. If we let $B = B_0 + C$, we have $B \in B_0 + \mathfrak{Q}_S$ and, using (3.40),

$$\|B - B_0\| = \|C\| \leqslant |\Psi - \Psi_0| \leqslant \frac{1}{\varepsilon}\left[P(B_0) - \sigma_0(B_0) - s(\sigma_0)\right],$$

which proves (3.38). Furthermore $\rho(C) = \rho(A_{\Psi-\Psi_0})$ for all $\rho \in I$, so that $I_B = I_{A_\Psi} \ni \sigma$, which concludes the proof.

3.20 Theorem

Let $A \in \mathfrak{C}_\Delta$ for some finite $\Delta \subset \mathbf{Z}^\nu$, and define a convex cone

$$\mathfrak{Q} = \left\{ aA + \sum_{x \in \mathbf{Z}^\nu} b_x A \cdot (A \circ \tau^x) : a, b_x \in \mathbf{R}, b_x \geqslant 0, \sum b_x < +\infty \right\}.$$

(a) *Let $\sigma_0', \sigma_0'' \in I$ be such that $\sigma_0'(A) \neq \sigma_0''(A)$. Given $C \in \mathfrak{C}$, there exist $B \in C + \mathfrak{Q}$ and two equilibrium states $\sigma', \sigma'' \in I_B$ such that $\sigma'(A) \neq \sigma''(A)$.*
(b) *Let σ_0', σ_0'' be equilibrium states for $C \in \mathfrak{C}$, such that $\sigma_0'(A) \neq \sigma_0''(A)$. Given $\varepsilon > 0$, one can choose $\delta > 0$ such that, if $C' \in \mathfrak{C}$ and $\|C' - C\| < \delta$, there exist $B \in C' + \mathfrak{Q}$ with $\|B - C'\| < \varepsilon$, and two equilibrium states $\sigma', \sigma'' \in I_B$ with $\sigma'(A) \neq \sigma''(A)$.*

Write $\sigma_0 = \frac{1}{2}(\sigma_0' + \sigma_0'')$. The assumption $\sigma_0'(A) \neq \sigma_0''(A)$ implies that

$$m_{\sigma_0}(\hat{A}^2) = \lim_{\Lambda \nearrow \infty} \sigma_0 \left[\left(|\Lambda|^{-1} \sum_{x \in \Lambda} A \circ \tau^x \right)^2 \right] > \sigma_0(A)^2$$

(see Section 3.6). Choose $\varepsilon > 0$ such that

$$\lim_{\Lambda \nearrow \infty} \sigma_0 \left[\left(|\Lambda|^{-1} \sum_{x \in \Lambda} A \circ \tau^x \right)^2 \right] \geqslant \sigma_0(A)^2 + 4\varepsilon \|A\|^2.$$

We apply Lemma 3.19 with $A_1 = A_2 = A$ and $S = \mathbf{Z}^\nu$ (we shall choose B_0 later), obtaining $B \in B_0 + \mathfrak{Q}$ and $\sigma \in I_B$ such that

$$\|B - B_0\| \leqslant \frac{1}{\varepsilon}\left[P(B_0) - \sigma_0(B_0) - s(\sigma_0)\right]$$

and

$$\sigma \left[\left(|\Lambda|^{-1} \sum_{x \in \Lambda} A \circ \tau^x \right)^2 \right] - \sigma(A)^2 \geqslant \sigma_0$$

$$\times \left[\left(|\Lambda|^{-1} \sum_{x \in \Lambda} A \circ \tau^x \right)^2 \right] - \sigma_0(A)^2 - 3\varepsilon \|A\|^2.$$

Therefore

$$m_\sigma(\hat{A}^2) = \lim_{\Lambda \nearrow \infty} \sigma\left[\left(|\Lambda|^{-1} \sum_{x\in\Lambda} A \circ \tau^x\right)^2\right] \geqslant \sigma(A)^2 + \varepsilon\|A\|^2.$$

From this it follows that there exist σ', σ'' in the support of m_σ with $\sigma'(A) \neq \sigma''(A)$. Taking $B_0 = C$ we obtain (a).

Suppose now that $\sigma_0', \sigma_0'' \in I_C$. Choose $\delta > 0$ such that if $C' \in \mathcal{C}$ and $\|C' - C\| < \delta$, we have

$$P(C') - \sigma_0(C') - s(\sigma_0) < \varepsilon^2.$$

Taking $B_0 = C'$, we have $B \in C' + \mathcal{Q}$, and (3.38) gives $\|B - C'\| < \varepsilon$, proving (b).

3.21 Coexistence of phases

In the above lemma and theorem we could restrict our attention to interactions Φ such that $\Phi(\xi|X) = 0$ when $|X| > 2|\Delta|$. Such interactions belong to the space $\mathcal{B}$ to be introduced in Chapter 4, and are "physically acceptable." Theorem 3.20 deals with the situation where there are at least two different equilibrium states. This corresponds physically to the coexistence of at least two phases. Part (b) of Theorem 3.20 shows that an interaction Ψ_0 (or a function C) for which several phases coexist cannot be isolated: it lies in an "infinite-dimensional manifold" of such interactions. One should check that these interactions are not all "physically equivalent" (see Section 4.7). For this see Exercise 2.

The coexistence of at least $n + 1$ phases can be treated in a similar manner. Let $A_1, \ldots, A_n \in \mathcal{C}_\Delta$, and let

$$A = \sum a_i A_i, \qquad \text{with} \quad \sum a_i^2 = 1.$$

We assume that $\sigma_0^{(0)}, \sigma_0^{(1)}, \ldots, \sigma_0^{(n)} \in I$ are such that

$$\sigma_0^{(0)}(A) = \sigma_0^{(1)}(A) = \cdots = \sigma_0^{(n)}(A)$$

does not hold for any choice of $a_1, \ldots, a_n$. Defining

$$\sigma_0 = \frac{1}{n+1} \sum_{i=0}^{n} \sigma_0^{(i)},$$

we have

$$m_{\sigma_0}(\hat{A}^2) - \sigma_0(A)^2 \geqslant 4\varepsilon\|A\|^2$$

with some $\varepsilon > 0$ independent of $a_1, \ldots, a_n$. Let $\mathcal{Q}$ be the linear space generated by the A_i and $A_i \cdot (A_j \circ \tau^x)$, and let $B_0 \in \mathcal{C}$. There exists $B \in B_0 + \mathcal{Q}$ such that

$$\|B - B_0\| \leqslant \frac{1}{\varepsilon}[P(B_0) - \sigma_0(B_0) - s(\sigma_0)],$$

and $\sigma \in I_B$ such that

$$|\sigma(A \cdot (A \circ \tau^x)) - \sigma(A)^2 - [\sigma_0(A \cdot (A \circ \tau^x)) - \sigma_0(A)^2]| \leqslant 3\varepsilon\|A\|^2$$

for all $a_1, \ldots, a_n$ and all $x \in \mathbf{Z}^\nu$. Therefore

$$m_\sigma(\hat{A}^2) - \sigma(A)^2 \geqslant \varepsilon\|A\|^2,$$

proving that the dimension of I_B is at least n: at least $n + 1$ phases coexist. Again an interaction for which at least $n + 1$ phases coexist cannot be isolated.

Notes

The statistical mechanics with translation invariance described in this chapter has been developed by physicists, mostly earlier than the theory of Gibbs states in Chapters 1 and 2. The thermodynamic limit for the pressure has been proved in various forms by various people, and the notion of equilibrium state has emerged progressively. Gallavotti and Miracle [1] noted the important fact (Theorem 3.7(b)) that for a residual set of interactions there is only one equilibrium state. For the discussion of the entropy (Sections 3.8 to 3.10) see Robinson and Ruelle [1]; for the variational principle (Theorem 3.12) see Ruelle [1]).* Sections 3.16 to 3.21 are based on the work of Israel [1], which has recently shed some light on the general nature of phase transitions.

Exercises

1. We write $|x| = \max_i |x_i|$ when $x \in \mathbf{Z}^\nu$. Let $0 < \lambda < 1$. If $\xi, \eta \in \Omega$, define

$$d(\xi, \eta) = \lambda^k \qquad \text{where} \qquad k = \inf\{|x| : \xi_x \neq \eta_x\}.$$

   (a) Check that d is a metric compatible with the topology of Ω.
   (b) Let $0 < \alpha < 1$. A function $A : \Omega \mapsto \mathbf{R}$ such that $|A(\xi) - A(\eta)| \leqslant c d(\xi, \eta)^\alpha$ for some $c \geqslant 0$ is said to be Hölder continuous of exponent α.

* In general, see Ruelle [3] for further details and references.

Such functions form a Banach space $\mathcal{C}^\alpha(\Omega)$ with the norm

$$\|A\|_\alpha = \max\left(\max_\xi |A(\xi)|, \sup_{\xi\neq\eta} \frac{|A(\xi) - A(\eta)|}{d(\xi,\eta)^\alpha}\right).$$

Write $\operatorname{diam} X = \max\{|y - x| : x, y \in X\}$, and let $\theta = \lambda^{\alpha/2}$. Show that if $A \in \mathcal{C}^\alpha(\Omega)$, then $A = A_\Phi$, with

$$\sup_X \theta^{-\operatorname{diam} X} \sup_{\xi\in\Omega} |\Phi(\xi|X)| < +\infty.$$

[For (b), let $\Lambda_n = \{x \in \mathbf{Z}^\nu : |x| \leqslant n\}$ and write $A = \sum_{n=0}^\infty A_n$, where $A_n \in \mathcal{C}_{\Lambda_n}$ and $\|A_n\| \leqslant \|A\|_\alpha (\lambda^n)^\alpha$. Proceed then as in Section 3.2.]

2. We consider a system ("lattice gas") with $\Omega_0 = \{0, 1\}$ and $\Omega = \{0, 1\}^{\mathbf{Z}^\nu}$. We define $A \in \mathcal{C}_{\{0\}}$ by $A(\xi) = \xi_0$ (A takes therefore the values 0 and 1). We shall use "pair" interactions Φ, such that $\Phi(\xi|X) = 0$ if $|X| > 2$ and

$$\Phi(\xi|\{0\}) = -\mu A(\xi),$$
$$\Phi(\xi|\{0, x\}) = \varphi(x)A(\xi)A(\tau^x\xi) \qquad \text{for} \qquad x \neq 0.$$

Here $\mu \in \mathbf{R}$, and $\varphi(x) = \varphi(-x) \in \mathbf{R}$ is defined for $x \neq 0$. Notice that

$$|\Phi| = |\mu| + \frac{1}{2}\sum_{x\neq 0} |\varphi(x)|.$$

(a) Let $0 \in M \subset \mathbf{Z}^\nu$, M finite, $M = -M$. Suppose that a function $\tilde{\varphi} : M \setminus \{0\} \mapsto \mathbf{R}$ is given such that $\tilde{\varphi}(x) = \tilde{\varphi}(-x)$. One can then extend $\tilde{\varphi}$ to $\varphi : \mathbf{Z}^\nu \setminus \{0\} \to \mathbf{R}$ such that

$$\sum_{x\neq 0} |\varphi(x)| < +\infty$$

and find μ such that there are two equilibrium states σ', σ'' for Φ satisfying $\sigma'(A) \neq \sigma''(A)$.

(b) Let μ_0, φ_0 correspond to a pair interaction Φ_0. We assume that σ_0', σ_0'' are equilibrium states for Φ_0, and that $\sigma_0'(A) \neq \sigma_0''(A)$. Given $\varepsilon > 0$, there exists δ such that the following holds.

Let $0 \in M \subset \mathbf{Z}^\nu$, M finite, $M = -M$. Suppose that $\tilde{\varphi} : M\setminus\{0\} \mapsto \mathbf{R}$ satisfies $\tilde{\varphi}(x) = \tilde{\varphi}(-x)$ and

$$\frac{1}{2}\sum_{x\in M\setminus\{0\}} |\tilde{\varphi}(x) - \varphi_0(x)| < \delta.$$

One can then extend $\tilde{\varphi}$ to $\varphi : \mathbf{Z}^\nu \setminus \{0\} \mapsto \mathbf{R}$ and find μ such that

$\varphi(x) = \varphi(-x)$, $\varphi(x) \leqslant \varphi_0(x)$ if $x \notin M$,

$$|\mu - \mu_0| + \frac{1}{2} \sum_{x \notin M} |\varphi(x) - \varphi_0(x)| < \varepsilon,$$

and there are two equilibrium states σ', σ'' for the interaction Φ corresponding to μ, φ, satisfying $\sigma'(A) \neq \sigma''(A)$. [Imitate the proof of Theorem 3.20, using Lemma 3.19 with $S = \mathbf{Z}^\nu \setminus M$. Notice that

$$m_\rho(\hat{A}^2) = \lim_{\Lambda \nearrow \infty} \rho \left[|\Lambda|^{-2} \sum_{\substack{x,y \in \Lambda \\ x-y \notin M}} (A \circ \tau^x) \cdot (A \circ \tau^y) \right].$$

Remark: It will be seen (Exercise 2 of Chapter 4) that Φ and Φ_0 are physically equivalent only if $\mu = \mu_0$ and $\varphi = \varphi_0$.]

4

Connection between Gibbs states and Equilibrium states

This chapter establishes the connection between the Gibbs states and the equilibrium states introduced earlier.

4.1 Generalities

In the present chapter we use the general assumptions of both Chapter 1 and Chapter 3. Thus a family $(\bar{\Omega}_\Lambda)_{\Lambda \in \mathcal{F}}$ is given, it is translation invariant, and if $S \subset \mathbf{Z}^\nu$ we write

$$\Omega_S = \left\{ \xi \in \prod_{x \in S} \Omega_x : (\forall \Lambda \in \mathcal{F} : \Lambda \subset S)\, \xi | \Lambda \in \bar{\Omega}_\Lambda \right\}.$$

We introduce also the Banach space $\mathcal{B}$ of translation invariant interactions with the norm

$$\Phi \mapsto \|\Phi\| = \sum_{X \ni 0} \sup_{\xi \in \Omega_X} |\Phi(\xi)| < +\infty;$$

clearly $\mathcal{A}_0 \subset \mathcal{B} \subset \mathcal{A}$. If we define $\varphi : \mathcal{B} \mapsto \mathcal{C}$ by $\varphi\Phi = A_\Phi, \varphi\mathcal{B}$ is dense in $\mathcal{C}$ and φ is continuous:

$$\|\Phi\| \geqslant |\Phi| \geqslant \|A_\Phi\|.$$

Therefore Theorem 3.7(c) and (d) apply with $\mathcal{X} = \mathcal{B}$.

If $\Phi \in \mathcal{B}$, then the set K_Φ of Gibbs states for Φ and the set $I_{\varphi\Phi}$ of equilibrium states for Φ are defined. We shall see that $I_{\varphi\Phi} \subset K_\Phi \cap I$. The reverse inclusion $I_{\varphi\Phi} \supset K_\Phi \cap I$ is not generally true, but holds if Ω satisfies the following condition.

(D) *There exist sequences* $(\Lambda_n), (M_n)$ *with* $\Lambda_n \nearrow \infty$, $\Lambda_n \subset M_n$, $|\Lambda_n|/|M_n| \to 1$, *and for each* $\xi, \eta \in \Omega$ *and* n *there exists* $\zeta_n \in \Omega$ *such that*

$$\zeta_n | \Lambda_n = \xi | \Lambda_n, \qquad \zeta_n | (\mathbf{Z}^\nu \setminus M_n) = \eta | (\mathbf{Z}^\nu \setminus M_n).$$

Notice that this condition is stronger than (D*) of Remark 1.14. (D) and (D*) reduce to "mixing" when $\nu = 1$ (see Chapter 5).

4.2 Theorem

If $\Phi \in \mathcal{B}$, then $I_{A_\Phi} \subset K_\Phi \cap I$. If furthermore Ω satisfies the above condition (D), *then $I_{A_\Phi} = K_\Phi \cap I$, i.e., an invariant state is an equilibrium state if and only if it is a Gibbs state.*

If $\Phi \in \mathcal{C}_0$, the Gibbs ensemble with boundary $\mu_{(\Lambda)_\eta}$ depends only on $\eta|(M \setminus \Lambda)$ for some finite set $M = \Lambda + \Delta$ (the set Δ depends on Φ, and also on $\mathcal{F}$). Using the definition (3.33) in the proof of Theorem 3.12, one checks easily that

$$\rho_\Lambda\{\xi\} = \sum_{\eta \in \Omega_{M\setminus\Lambda}} \mu_{(\Lambda)_\eta}\{\xi\}\rho_{M\setminus\Lambda}\{\eta\}.$$

The state ρ^Φ defined by $\rho_\Lambda = \alpha_\Lambda \rho^\Phi$ is thus a Gibbs state. On the other hand the proof of Theorem 3.12 shows that $\rho^\Phi \in I_{\varphi\Phi}$. We have thus established the existence of $\rho^\Phi \in I_{\varphi\Phi} \cap K_\Phi$ when $\Phi \in \mathcal{C}_0$.

Let now $\Phi \in \varphi^{-1}D$, $\Phi = \lim_{n\to\infty} \Phi_n$, $\Phi_n \in \mathcal{C}_0$. Any limit of ρ^{Φ_n} when $n \to \infty$ is in $I_{\varphi\Phi} = \{\rho_\Phi\}$ and also in K_Φ, as one sees immediately. Therefore there exists $\rho^\Phi \in I_{\varphi\Phi} \cap K_\Phi$ if $\Phi \in \varphi^{-1}D$.

For an arbitrary $\Phi \in \mathcal{B}$ we see, applying Theorem 3.7(d), that $I_{\varphi\Phi}$ is in the closed convex hull of states $\rho \in K_\Phi \cap I$. Therefore $I_{\varphi\Phi} \subset K_\Phi \cap I$.

We assume now that Ω satisfies (D) and have to show that if $\sigma \in K_\Phi \cap I$, then $\sigma \in I_{\varphi\Phi}$. Somewhat more generally we shall prove that if $\sigma \in K_\Phi$, then

$$\liminf_{n\to\infty} |M_n|^{-1} \left[S(\alpha_{M_n}\sigma) - (\alpha_{M_n}\sigma)\left(U^\Phi_{M_n}\right)\right] \geqslant P^\Phi.$$

By the definition of a Gibbs state,

$$(\alpha_{M_n}\sigma)\{\xi\} = \int_{\Omega_{\mathbf{Z}^\nu\setminus M_n}} \sigma_{\mathbf{Z}^\nu\setminus M_n}(d\eta)\mu_{(M_n)\eta}\{\xi\},$$

and by the concavity of the entropy,

$$\begin{aligned} S(\alpha_{M_n}\sigma) - (\alpha_{M_n}\sigma)\left(U^\Phi_{M_n}\right) &\geqslant \int_{\Omega_{\mathbf{Z}^\nu\setminus M_n}} \sigma_{\mathbf{Z}^\nu\setminus M_n}(d\eta)\left[S(\mu_{(M_n)\eta}) - \mu_{(M_n)\eta}\left(U^\Phi_{M_n}\right)\right] \\ &= \int_{\Omega_{\mathbf{Z}^\nu\setminus M_n}} \sigma_{\mathbf{Z}^\nu\setminus M_n}(d\eta) \sum_{\xi\in\Omega_{M_n}} \mu_{(M_n)\eta}\{\xi\} \end{aligned}$$

$$\times \left[W_{M_n,\mathbf{Z}^\nu\setminus M_n}(\xi \vee \eta) + \log \sum_{\zeta\in\Omega_{M_n}} \right.$$

$$\left. \exp[-U_{M_n}(\zeta) - W_{M_n,\mathbf{Z}^\nu\setminus M_n}(\zeta \vee \eta)] \right].$$

Thus

$$\begin{aligned}
&\liminf_{n\to\infty} |M_n|^{-1}\big[S(\alpha_{M_n}\sigma) - (\alpha_{M_n}\sigma)\big(U^\Phi_{M_n}\big)\big] \\
&\quad \geqslant \liminf_{n\to\infty} |M_n|^{-1} \int_{\Omega_{\mathbf{Z}^\nu\setminus M_n}} \sigma_{\mathbf{Z}^\nu\setminus M_n}(d\eta) \\
&\quad \times \log \sum_{\zeta\in\Omega_{M_n}} \exp[-U_{M_n}(\zeta) - W_{M_n,\mathbf{Z}^\nu\setminus M_n}(\zeta \vee \eta)].
\end{aligned}$$

In view of (D) we have (when $\eta = \eta^*|(\mathbf{Z}^\nu\setminus M_n)$ for some $\eta^* \in \Omega$)

$$\sum_{\zeta\in\Omega_{M_n}} \exp[-U_{M_n}(\zeta) - W_{M_n,\mathbf{Z}^\nu\setminus M_n}(\zeta \vee \eta)] \geqslant \sum_{\xi\in\Omega^*_{\Lambda_n}} \exp[-U_{\Lambda_n}(\xi) + R_n],$$

where $\lim_{n\to\infty} |R_n|/|M_n| = 0$; hence

$$\begin{aligned}
&\liminf_{n\to\infty} |M_n|^{-1}\big[S(\alpha_{M_n}\sigma) - (\alpha_{M_n}\sigma)\big(U^\Phi_{M_n}\big)\big] \\
&\quad \geqslant \lim_{n\to\infty} |M_n|^{-1} \log \sum_{\xi\in\Omega^*_{\Lambda_n}} \exp[-U_{\Lambda_n}(\xi) + R_n] = P^\Phi.
\end{aligned}$$

4.3 Physical interpretation

When a pure thermodynamic phase for the interaction $\Phi \in \mathcal{B}$ has a non-trivial decomposition into pure Gibbs states, we say that there is *symmetry breakdown*, or more precisely that the translation invariance of the theory is broken.

4.4 Proposition

Let condition (D) hold. If $\sigma \in K_\Phi$, $\Phi \in \mathcal{B}$, and $C \in \mathcal{C}$, then

$$\lim_{n\to\infty} \frac{1}{|\Lambda_n|} \log \sigma \left(\exp \sum_{x\in\Lambda_n} C \circ \tau^x \right) = P(A_\Phi + C) - P(A_\Phi).$$

By density and equicontinuity it suffices to consider the case $C = A_\Psi$, $\Psi \in \mathcal{A}_0$. We then have

$$\lim_{n\to\infty} \frac{1}{|\Lambda_n|} \log \sigma \left(\exp \sum_{x\in\Lambda_n} C \circ \tau^x \right) = \lim_{n\to\infty} \frac{1}{|\Lambda_n|} \log(\alpha_{\Lambda_n}\sigma)\big(\exp\big(-U^\Psi_{\Lambda_n}\big)\big).$$

If $\xi, \eta \in \Omega$, it follows from (D) that

$$\frac{[(\alpha_{\Lambda_n}\sigma)\{\xi|\Lambda_n\}]\exp U^{\Phi}_{\Lambda_n}(\xi|\Lambda_n)}{[(\alpha_{\Lambda_n}\sigma)\{\eta|\Lambda_n\}]\exp U^{\Phi}_{\Lambda_n}(\eta|\Lambda_n)} \leqslant e^{R_n},$$

where $R_n/|\Lambda_n| \to 0$ when $n \to \infty$. Therefore

$$e^{-R_n}\mu_{(\Lambda_n)}\{\xi_{\Lambda_n}\} \leqslant (\alpha_{\Lambda_n}\sigma)(\xi_{\Lambda_n}) \leqslant e^{R_n}\mu_{(\Lambda_n)}\{\xi_{\Lambda_n}\};$$

hence

$$e^{-R_n} Z^{\Phi+\Psi}_{\Lambda_n}/Z^{\Phi}_{\Lambda_n} \leqslant (\alpha_{\Lambda_n}\sigma)\big(\exp\big(-U^{\Psi}_{\Lambda_n}\big)\big) \leqslant e^{R_n} Z^{\Phi+\Psi}_{\Lambda_n}/Z^{\Phi}_{\Lambda_n}$$

and the proposition follows.

4.5 Remark

Let τ be an action of $\mathbf{Z}^\nu$ by homeomorphisms of a metrizable compact space Ω, σ_0 a τ-invariant probability measure, and A a continuous real function on Ω. For finite $\Lambda \subset \mathbf{Z}^\nu$, define

$$Z^*_\Lambda = \sigma_0\left(\exp\sum_{a\in\Lambda} A\circ\tau^a\right).$$

Sinai [4] has proposed (for $\nu = 1$) to call a Gibbs state any limit as $\Lambda \to \mathbf{Z}^\nu$ of

$$Z^{*-1}_\Lambda\left[\exp\sum_{a\in\Lambda} A(\tau^a\xi)\right]\sigma_0(d\xi).$$

A result relevant to this definition is as follows.

Let $\Phi, \Psi \in \mathcal{B}$ and $\sigma \in K_\Phi$. For finite $\Lambda \subset \mathbf{Z}^\nu$, and $\xi \in \Omega_\Lambda$, define

$$Z^*_\Lambda = \sigma\left(\exp\sum_{a\in\Lambda} A_\Psi\circ\tau^a\right).$$

Then any limit as $\Lambda \to \mathbf{Z}^\nu$ of

$$Z^{*-1}_\Lambda\left[\exp\sum_{a\in\Lambda} A_\Psi\circ\tau^a\right]\sigma$$

belongs to $K_{\Phi+\Psi}$.

[This can be proved easily by the methods of Chapter 1; see Exercise 4 of that chapter.]

4.6 Strict convexity of the pressure

We know by Theorem 3.4 that the function $\Phi \to P^{\Phi} = P(A_{\Phi})$ is convex and continuous on $\mathcal{B}$. Suppose that the graph of this function contains a straight-line interval, i.e. there exist $\Phi, \Psi \in \mathcal{B}$, $\Psi \neq 0$, such that

$$P^{\Phi+t\Psi} = P^{\Phi} + ct \qquad \text{for} \quad t \in [-1, +1].$$

By Theorem 3.7, if ρ is an equilibrium state for Φ, it is also an equilibrium state for $\Phi + t\Psi$ when $|t| \leqslant 1$. Therefore ρ is a Gibbs state for $\Phi + t\Psi$ when $|t| \leqslant 1$. Since the expression

$$\int_{\Omega_{L\backslash\Lambda}} \rho_{L\backslash\Lambda}(d\eta)\mu_{(\Lambda)\eta}^{\Phi+t\Psi}\{\xi_{\Lambda}\}$$

is real analytic in t and constant for $|t| \leqslant 1$, it is constant for all real t. Hence ρ is a Gibbs state for $\Phi + t\Psi$, all real t. If condition (D) is satisfied, Theorem 4.2 shows that ρ is also an equilibrium state and thus

$$\max_{\sigma \in I}[s(\sigma) + \sigma(A_{\Phi}) + t\sigma(A_{\Psi})] = s(\rho) + \rho(A_{\Phi+t\Psi}) = P^{\Phi} + ct$$

for all $t \in \mathbf{R}$. This implies $\sigma(A_{\Psi}) = c$ for all $\sigma \in I$, or $\sigma(A_{\Psi} - c) = 0$ for all translation invariant measures σ on Ω. Therefore $A_{\Psi} - c$ belongs to the closed subspace $\mathcal{J}$ of $\mathcal{C}$ generated by the elements $A \circ \tau^{a} - A$ with $A \in \mathcal{C}, a \in \mathbf{Z}^{\nu}$. Summarizing, we have proved the following result.

4.7 Proposition

Let $\varphi : \mathcal{B} \mapsto \mathcal{C}$ be defined by $\varphi\Phi = A_{\Phi}$, and let $\mathcal{J}$ be the closed subspace of $\mathcal{C}$ generated by the elements $A \circ \tau^{a} - A$, with $A \in \mathcal{C}, a \in \mathbf{Z}^{\nu}$. We denote by $[\Phi]$ the image of $\Phi \in \mathcal{B}$ in the quotient $\varphi\mathcal{B}/(\varphi\mathcal{B} \cap \mathcal{J})$.

(a) *The function $[\Phi] \mapsto P(A_{\Phi})$ is well defined on $\varphi\mathcal{B}/(\varphi\mathcal{B} \cap \mathcal{J})$, and if condition (D) holds, this function is strictly convex on the subset $\{[\Phi] \in \varphi\mathcal{B}/(\varphi\mathcal{B} \cap \mathcal{J}) : \sigma(A_{\Phi}) = 0\}$, where σ is some arbitrarily chosen element of I.*
(b) *If (D) holds and if ρ is an equilibrium state for both $\Phi, \Phi' \in \mathcal{B}$, then $A_{\Phi'} - A_{\Phi} \in \mathcal{J} + \mathbf{R}$.*

Let us say that $\Phi, \Phi' \in \mathcal{B}$ are *physically equivalent* if there are $c \in \mathbf{R}$ and $B \in \mathcal{J}$ such that $A_{\Phi'-\Phi} = B + c$ (or in other notation $[\Phi'] = [\Phi] + c$ with $c \in \mathbf{R}$). Two physically equivalent interactions in $\mathcal{B}$ then have the same equilibrium states. Conversely, if two interactions in $\mathcal{B}$ have an equilibrium state in common, they are physically equivalent. The restriction of P to an equivalence class

$\{\Phi' \in \mathcal{B} : A_{\Phi'-\Phi} \in \mathcal{I} + \mathbf{R}\}$ is affine, and the equivalence classes are maximal affine sets on which P is affine.

4.8 $\mathbf{Z}^\nu$-lattice systems and $\mathbf{Z}^\nu$-morphisms

In Section 2.1 we have introduced objects $(L, (\Omega_x)_{x\in L}, (\bar{\Omega}_\Lambda)_{\Lambda\in\mathcal{F}})$ called lattice systems. In the present chapter we have assumed that $L = \mathbf{Z}^\nu$, $\Omega_x = \Omega_0$, and $(\bar{\Omega}_\Lambda)_{\Lambda\in\mathcal{F}}$ is invariant under translations of $\mathbf{Z}^\nu$. Such a lattice system with the added structure provided by the translations will be called a $\mathbf{Z}^\nu$*-lattice system*, and will be denoted by $\left(\mathbf{Z}^\nu, \Omega_0, (\bar{\Omega}_\Lambda)_{\Lambda\in\mathcal{F}}\right)$. We say that

$$F : \left(\mathbf{Z}^\nu, \Omega'_0, (\bar{\Omega}'_{\Lambda'})_{\Lambda'\in\mathcal{F}}\right) \mapsto \left(\mathbf{Z}^\nu, \Omega_0, (\bar{\Omega}_\Lambda)_{\Lambda\in\mathcal{F}}\right)$$

is a $\mathbf{Z}^\nu$*-morphism* if a family $(F_x)_{x\in\mathbf{Z}^\nu}$ exists with the properties (M1)–(M4) of Section 2.1 and furthermore

(M5) $$F_{x-a}\tau^a = F_x.$$

In particular, F is a morphism. If it is an isomorphism we say that F is a $\mathbf{Z}^\nu$*-isomorphism*.

4.9 Proposition

A map $F : \Omega' \mapsto \Omega$ is a $\mathbf{Z}^\nu$-morphism if and only if it has the following properties:

(a) *F is continuous;*
(b) *F is equivariant, i.e. $\tau^a F = F\tau^a$;*
(c) *F restricted to $\sum'_{\xi'} = \{\eta' \in \Omega' : lim_{x\to\infty} d'(\tau^x\eta', \tau^x\xi') = 0\}$ is a bijection to $\sum_{F\xi'} = \{\eta \in \Omega : lim_{x\to\infty} d(\tau^x\eta, \tau^x F\xi') = 0\}$.* [Here d, d' are distance compatible with the topologies of Ω, Ω' respectively.]

(M1)–(M5) clearly imply (a)–(c). Suppose now that (a), (b), and (c) hold. Since F is continuous, there is a finite $M \subset \mathbf{Z}^\nu$ such that $(F\xi')_0$ depends only on $\xi'|M$. Define $M' = M \bigcup \cup\{M + \Lambda : 0 \in \Lambda \in \mathcal{F}\}$. By compactness one can choose a finite $M'' \supset M'$ such that for all $\eta' \in \Omega'_{M''}$ there exists $\eta^* \in \Omega'$ satisfying

$$\eta^*|M' = \eta'|M'.$$

Define $M(x) = M'' + x$, and $F_x : \Omega'_{M(x)} \mapsto \Omega_x$ by

$$F_x(\tau^{-x}\eta') = (F\eta^*)_0.$$

(M1), (M2), (M4), and (M5) are clearly satisfied. If $\Lambda \in \mathcal{F}$, $\Lambda \subset X$, and $\xi' \in \Omega'_{\cup\{M(x):x\in X\}}$, take $x_0 \in \Lambda$ and let $\eta' = \tau^{x_0}\xi'|M''$. There exists $\eta^* \in \Omega'$ such that $\eta^*|M' = \eta'|M' = \tau^{x_0}\xi'|M'$; therefore, if $x \in \Lambda$, we have $\eta^*|M + (x - x_0) = \tau^{x_0}\xi'|M + (x - x_0)$ by definition of M'. Notice that

$$\begin{aligned} F_x(\xi'|M(x)) &= F_x(\tau^{-x}\tau^{x-x_0}(\tau^{x_0}\xi'|M'' + (x - x_0))) \\ &= F_x(\tau^{-x}\tau^{x-x_0}(\eta^*|M'' + (x - x_0))) \\ &= (F\tau^{x-x_0}\eta^*)_0 = (F\eta^*)_{x-x_0}, \end{aligned}$$

and therefore (M3) holds.

4.10 Corollary

(a) *If $F : \Omega' \mapsto \Omega$ is an equivariant homeomorphism, then F is a $\mathbf{Z}^\nu$-isomorphism.*

(b) *For each $\mathbf{Z}^\nu$-lattice system $(\mathbf{Z}^\nu, \Omega_0, (\bar{\Omega}_\Lambda)_{\Lambda\in\mathcal{F}})$ there is a $\mathbf{Z}^\nu$-isomorphism*

$$F : (\mathbf{Z}^\nu, \Omega'_0, (\bar{\Omega}'_{\Lambda'})_{\Lambda'\in\mathcal{F}'}) \mapsto (\mathbf{Z}^\nu, \Omega_0, (\bar{\Omega}_\Lambda)_{\Lambda\in\mathcal{F}}),$$

where $\mathcal{F}'$ consists of the two-point sets $\{x, y\}$ such that x and y are nearest neighbors, i.e. $\sum_{i=1}^{\nu} |x_i - y_i| = 1$.

The proof of (a) is immediate.

To prove (b) let $M(x) = \{y \in \mathbf{Z}^\nu : max_i|x_i - y_i| \leqslant l\}$, where $l \geqslant 0$ is such that if $x \in \Lambda \in \mathcal{F}$, then $\Lambda \subset M(x)$. Let also $\Omega'_x = \Omega_{M(x)}$. As announced, we take

$$\mathcal{F}' = \{\{x, y\}\text{:x and y are nearest neighbors}\}.$$

If $\{x, y\} \in \mathcal{F}'$ we let

$$\bar{\Omega}'_{\{x,y\}} = \{(\xi, \eta) \in \Omega_{M(x)} \times \Omega_{M(y)} : \xi|M(x) \cap M(y) = \eta|M(x) \cap M(y)\}.$$

The map $F : \Omega' \mapsto \Omega$ such that $(F\xi')_x = (\xi'_x)_x$ is an equivariant homeomorphism; hence, by (a), it is a $\mathbf{Z}^\nu$-isomorphism.

4.11 Remark

The results of Chapter 2 apply immediately to $\mathbf{Z}^\nu$-morphisms and translationally invariant interactions. In particular if $\Phi \in \mathcal{B}(\mathbf{Z}^\nu, \Omega_0, (\bar{\Omega}_\Lambda)_{\Lambda\in\mathcal{F}})$, and F is a $\mathbf{Z}^\nu$-morphism, then $F^*\Phi \in \mathcal{B}(\mathbf{Z}^\nu, \Omega'_0, (\bar{\Omega}'_{\Lambda'})_{\Lambda'\in\mathcal{F}'})$, as follows from the norm estimate in Section 2.3.

4.12 Proposition

Let $F : (\mathbf{Z}^\nu, \Omega'_0, (\bar{\Omega}'_{\Lambda'})_{\Lambda' \in \mathcal{F}}) \mapsto (\mathbf{Z}^\nu, \Omega_0, (\bar{\Omega}_\Lambda)_{\Lambda \in \mathcal{F}})$ *be a* $\mathbf{Z}^\nu$*-morphism.*

(a) *If* σ' *is any* τ*-invariant state on* Ω'*, then* $s(\sigma') \geqslant s(F\sigma')$*, and*

$$\sigma'(A_{F^*\Phi}) = (F\sigma')(A_\Phi).$$

(b) *If* F *is a* $\mathbf{Z}^\nu$*-isomorphism, then* $P^{F^*\Phi} = P^\Phi$.

We have $S(\sigma'_{\cup\{M(x):x\in\Lambda\}}) \geqslant S((F\sigma')_\Lambda)$; hence $s(\sigma') \geqslant s(F\sigma')$. Furthermore

$$\begin{aligned}
\sigma'(A_{F^*\Phi}) &= -\sum_{X'\ni 0} \frac{1}{|X'|}\sigma'((F^*\Phi)\circ\alpha_{X'}) \\
&= -\sum_{X'\ni 0}\ \sum_{X:\cup\{M(x):x\in X\}=X'} \frac{1}{|X'|}\sigma'(\Phi\circ\alpha_X\circ F) \\
&= -\sum_{X:\cup\{M(x):x\in X\}\ni 0} \frac{1}{|\cup\{M(x):x\in X\}|}(F\sigma')(\Phi\circ\alpha_x) \\
&= -\sum_{X\ni 0}\frac{1}{|X|}(F\sigma')(\Phi\circ\alpha_X) \\
&= (F\sigma')(A_\Phi).
\end{aligned}$$

This proves (a); (b) follows from (a) and the variational principle for P (Theorem 3.12).

4.13 Restriction of $\mathbf{Z}^\nu$ to a subgroup G

Let G be a subgroup of finite index of $\mathbf{Z}^\nu$ (hence G is isomorphic to $\mathbf{Z}^\nu$) and let $(\mathbf{Z}^\nu, \Omega'_0, (\bar{\Omega}'_{\Lambda'})_{\Lambda'\in\mathcal{F}})$ be a $\mathbf{Z}^\nu$-lattice system. Choose $M(0) \subset \mathbf{Z}^\nu$ to contain exactly one element of each residue class of $\mathbf{Z}^\nu$ mod G, and let $M(x) = M(0) + x$ for each $x \in G$. *The family* $(M(x))_{x\in G}$ is a partition of $\mathbf{Z}^\nu$, and the construction of Example 2.2 yields an isomorphism $F : (\mathbf{Z}^\nu, \Omega'_0, (\bar{\Omega}'_{\Lambda'})_{\Lambda'\in\mathcal{F}}) \mapsto (G, \Omega'_{M(0)}, (\bar{\Omega}_\Lambda)_{\Lambda\in\mathcal{F}})$. (Note that this isomorphism is not a $\mathbf{Z}^\nu$-morphism if $G \neq \mathbf{Z}^\nu$.) We shall say that the G-lattice system $(G, \Omega'_{M(0)}, (\bar{\Omega}_\Lambda)_{\Lambda\in\mathcal{F}})$ is obtained from $(\mathbf{Z}^\nu, \Omega'_0, (\bar{\Omega}'_{\Lambda'})_{\Lambda'\in\mathcal{F}})$ by restriction of $\mathbf{Z}^\nu$ to the subgroup G. This definition is unique only up to G-isomorphism, because of the arbitrariness of the choice of $M(0)$.

4.14 Proposition

With the above notation, let $\Phi' \in \mathcal{B}\,(\mathbf{Z}^\nu, \Omega'_0, (\bar{\Omega}'_{\Lambda'})_{\Lambda' \in \mathcal{F}})$ *and* $\Phi'^* = (F^{-1})^*\Phi'$ *(we define* $(F^{-1})_x$ *to be the restriction map* $\Omega'_{M(a)} \mapsto \Omega'_x$ *when* $x \in M(a)$*). Then* $\Phi'^* \in \mathcal{B}(G, \Omega'_{M(0)}, (\bar{\Omega}_\Lambda)_{\Lambda \in \mathcal{F}})$.

(a) *If* σ *is any* τ_G*-invariant state on* Ω*, then*

$$\sigma(A_{\Phi'*} \circ F) = \sum_{x \in M(0)} \tau^x \sigma(A_{\Phi'}).$$

(b) *The pressure for the interaction* Φ'^* *and the group* G *is*

$$P^{\Phi'*} = |M(0)|\,P^{\Phi'}.$$

Similarly, if A *is real continuous on* Ω'

$$P\left(\sum_{x \in M(0)} A \circ \tau^x \circ F^{-1}\right) = |M(0)|\,P(A).$$

That $\Phi'^* \in \mathcal{B}$ follows from the norm estimate in Section 2.3 and the fact that Φ'^* is clearly G-invariant.

We have

$$\begin{aligned} A_{\Phi'^*}(F\xi) &= -\sum_{x \subset G}{}^*((F^{-1})^*\Phi')(F\xi|X) \\ &= -\sum_{x \subset G}{}^* \sum_{X' : \{a \in G : X' \cap M(a) \neq \emptyset\} = X} \Phi'(\xi|X'). \end{aligned}$$

For each finite $Y' \subset \mathbf{Z}^\nu$ and $x \in M(0)$, exactly one translate X' of $Y' + x$ by an element of G occurs in the sum on the right-hand side. This proves (a).

To prove (b) we apply the variational principle for P (Theorem 3.12), making use of (a) and of the following easily verified facts:

(i) the entropy of σ with respect to G is equal to the entropy of $\tau^x\sigma$, and therefore to the entropy of $\sigma' = |M(0)|^{-1}\Sigma_{x \in M(0)}\tau^x\sigma$ with respect to G,
(ii) the entropy of σ' with respect to G is $|M(0)|$ times its entropy with respect to $\mathbf{Z}^\nu$.

4.15 Undecidability and non-periodicity

The following curious facts are known about $\mathbf{Z}^2$-lattice systems (for a readable account of (a)–(c), see Robinson [1]).

(a) Let $\mathcal{F}$ denote the set of nearest-neighbor two-point subsets of $\mathbf{Z}^2$ (see Corollary 4.10(b)). Consider the following data: a finite set Ω_0, and a translation invariant family $(\bar{\Omega}_\Lambda)_{\Lambda\in\mathcal{F}'}$ such that $\bar{\Omega}_\Lambda \subset (\Omega_0)^\Lambda$. The problem of finding when these data determine a non-empty configuration space

$$\Omega = \{\xi \in (\Omega_0)^{\mathbf{Z}^2} : (\forall\Lambda \in \mathcal{F}')\, \xi|\Lambda \in \bar{\Omega}_\Lambda\}$$

is undecidable (Berger [1]).

(b) There is a $\mathbf{Z}^2$-lattice system as in (a) with no periodic configuration, i.e., if $\xi \in \Omega$ and $\tau^a\xi = \xi$, then $a = 0$. An example with $|\Omega_0| = 56$ and zero topological entropy (see Section 6.20) has been constructed by Robinson [1]. The existence of a $\mathbf{Z}^2$-lattice system with no periodic configuration is an ingredient in the proof of the undecidability property of (a), and is in turn implied by this property.

(c) Robinson [1] has constructed a system as in (a), with $|\Omega_0| = 36$, for which the following completion problem is undecidable: for an arbitrary finite $\Lambda \subset \mathbf{Z}^2$ and $\xi \in \Omega_\Lambda$, find if there exists $\xi^* \in \Omega$ with $\xi^*|\Lambda = \xi$.

(d) Let $K(l) = \{x \in \mathbf{Z}^2 : |x_1| \leqslant l, |x_2| \leqslant l\}$. Using the $\mathbf{Z}^2$-lattice system of (c), we define F to be the smallest function from the positive integers to the positive integers with the following property. An element $\xi \in \Omega_{k(l)}$ can be extended to some $\xi^* \in \Omega$ if and only if it can be extended to some $\xi^{**} \in \Omega_{K(F(l))}$. Since the completion problem in (c) is undecidable, F is not a computable function. In particular it grows extremely fast. This shows that, while the conditions $\xi|\Lambda \in \bar{\Omega}_\Lambda$ extend only over nearest-neighbor pairs $\Lambda \in \mathcal{F}'$, the effect of these conditions propagates over extremely large distances.

Notes

Dobrushin [2] showed that translation invariant Gibbs states are equilibrium states. The converse is due to Lanford and Ruelle [1]. The equivalence of the two notions is the main result of this chapter, and is of central importance in statistical mechanics. As an application, a theorem of "strict convexity of the pressure" is proved, following Griffiths and Ruelle [1]. The rest of the chapter is devoted to general results on morphisms compatible with the action of $\mathbf{Z}^\nu$.

Exercises

1. Let $A_1, A_2 \in \mathcal{C}_\Delta$, Δ finite, and suppose that, for some $\sigma_0 \in I$, $\sigma_0[A_1 \cdot (A_2 \circ \tau^x)]$ does not have a limit when $x \to \infty$. There is then an ergodic

equilibrium state σ for some interaction $\Phi \in \mathcal{B}$, such that $\sigma[A_1 \cdot (A_2 \circ \tau^x)]$ does not have a limit when $x \to \infty$. In particular σ is not a pure Gibbs state for Φ. [See Israel [1], Theorem 5. Define

$$\mathcal{Q} = \left\{ \frac{1}{2} \sum_{x \in \mathbf{Z}^\nu} (b_x A_1 \cdot (A_2 \circ \tau^x) + b_{-x}(A_1 \circ \tau^x) \cdot A_2) \right.$$
$$\left. : b_x \in \mathbf{R}, \sum_{x \in S} |b_x| < +\infty \right\}$$

and proceed as in the proof of Lemma 3.19. This gives σ' (not necessarily ergodic) with the desired property. There is also a state σ in the ergodic decomposition of σ' with the property, and Theorem 1.11 implies that σ is not a pure Gibbs state.]

2. We assume that $\mathcal{F} = \emptyset$, i.e. $\Omega = (\Omega_0)^{\mathbf{Z}^\nu}$. Let $\mathcal{S}_0 \subset \mathcal{B}$ be the subspace of interactions Φ such that

$$\sum_{\xi \in \Omega_\Lambda} \Phi(\xi \vee \eta) = 0$$

for all pairs of disjoint finite sets Λ, M, and all $\eta \in \Omega_M$. If $\Omega_0 = \{0, 1\}$, let $\mathcal{S}_1 \subset \mathcal{B}$ be the subspace of interactions Φ such that $\Phi(\xi|X) = 0$ unless $\xi_x = 1$ for all $x \in \Lambda$. Prove the following statements for $\mathcal{S} = \mathcal{S}_0$ and for $\mathcal{S} = \mathcal{S}_1$.

(a) If σ is a Gibbs state for both Φ and Φ', where, $\Phi', \Phi \in \mathcal{S}$, then $\Phi = \Phi'$.
(b) If Φ and Φ' are physically equivalent, and $\Phi, \Phi' \in \mathcal{S}$, then $\Phi = \Phi'$.
(c) The restriction of P to $\mathcal{S}$ is strictly convex.

[See Griffiths and Ruelle [1]. To prove (a) observe that, by Remark 1.14, $\operatorname{supp} \sigma = \Omega$. Since σ is a Gibbs state, the ratios

$$\frac{\exp[-U_\Lambda(\xi_\Lambda) - W_{\Lambda, \mathbf{Z}^\nu \setminus \Lambda}(\xi_\Lambda \vee \eta)]}{\exp[-U_\Lambda(\xi'_\Lambda) - W_{\Lambda, \mathbf{Z}^\nu \setminus \Lambda}(\xi'_\Lambda \vee \eta)]}$$

(with $\xi, \xi' \in \Omega_\Lambda$, $\eta \in \Omega_{\mathbf{Z}^\nu \setminus \Lambda}$) are the same for Φ and Φ'. In the case of $\mathcal{S}_0$, define "partial traces" $T_x : \mathcal{C} \to \mathcal{C}_x$ for X finite by

$$(T_x A)(\xi) = \lim_{M \to \mathbf{Z}^\nu} |\Omega_{M \setminus X}|^{-1} \sum_{\eta \in \Omega_{M \setminus X}} A((\xi|X) \vee \eta).$$

Using T_Λ, one see that $U_\Lambda(\xi_\Lambda) - U_\Lambda(\xi'_\Lambda)$ is the same for Φ, Φ'. One obtains $\Phi|\Omega_X = \Phi'|\Omega_X$ by induction on $|X|$. In the case of $\mathcal{S}_1$, take η such that $\eta_x = 0$ for all $x \in \mathbf{Z}^\nu \setminus \Lambda$. This gives that $U_\Lambda(\xi_\Lambda) - U_\Lambda(\xi'_\Lambda)$ is the same for

Φ, Φ', and one obtains again $\Phi|\Omega_X = \Phi'|\Omega_X$ by induction on $|X|$. For (b) and (c), see Section 4.7.]

3. Let $\Omega_0 = \{0, 1\}$ and assume that if, for some finite $\Lambda \subset \mathbf{Z}^\nu$, $\xi \in (\Omega_0)^{\mathbf{Z}^\nu}$ satisfies $\xi|\Lambda \in \Omega_\Lambda$ and $\xi_x = 0$ for $x \notin \Lambda$, then $\xi \in \Omega$. Show that the results of Exercise 2, with $\mathbb{S} = \mathbb{S}_1$, extend to such a $\mathbf{Z}^\nu$-lattice system (called "lattice gas with hard cores").

5

One-dimensional systems

The theory of Gibbs states on a "lattice" $\mathbf{Z}^\nu$ is best understood in the one-dimensional case, $\nu = 1$, which we shall now study. A **Z**-*lattice system* $(\mathbf{Z}, \Omega_0, (\bar{\Omega}_\Lambda)_{\Lambda\in\mathcal{F}})$ is also called a *subshift of finite type*. Notice that here τ^a may be interpreted as the ath power of the "shift" $\tau = \tau^1$.

Going over to a **Z**-isomorphic system (see Corollary 4.10(b)), we may assume that $\mathcal{F}$ consists of the two-point sets $\{x, x+1\}$. Let t be the square matrix indexed by $\Omega_0 \times \Omega_0$, with elements

$$t_{ij} = \begin{cases} 1 & \text{if} \quad (i, j) \in \bar{\Omega}_{\{x,x+1\}}, \\ 0 & \text{if} \quad (i, j) \notin \bar{\Omega}_{\{x,x+1\}}. \end{cases}$$

Then $\xi \in (\Omega_0)^{\mathbf{Z}}$ is an element of Ω if and only if

$$t_{\xi_x, \xi_{x+1}} = 1 \qquad \text{for all} \quad x \in \mathbf{Z}.$$

In what follows it will be convenient to use the notation (Ω_0, t) for this **Z**-lattice system. We shall also assume, as we may, that for each $i \in \Omega_0$ there exists $\xi \in \Omega$ such that $\xi_0 = i$.

We say that a **Z**-lattice system is *transitive* if the action of τ on Ω is topologically + transitive* – i.e., given two non-empty open sets $U, V \subset \Omega$, and $N \geqslant 0$, there exists $n > N$ such that $U \cap \tau^n V \neq \emptyset$. For (Ω_0, t) to be transitive it is necessary and sufficient that, for all $i, j \in \Omega_0$, one can choose an integer $a > 0$ such that $(t^a)_{ij} > 0$.

We say that a **Z**-lattice system is *mixing* if the action of τ on Ω is topologically mixing† – i.e., given two non-empty open sets $U, V \subset \Omega$ there exists $N \geqslant 0$ such that $U \cap \tau^n V \neq \emptyset$ for all $n > N$. For (Ω_0, t) to be mixing it is necessary

* See Appendix A.2.
† See Appendix A.2.

and sufficient that one can choose an integer $a > 0$ such that $(t^a)_{ij} > 0$ for all $i, j \in \Omega_0$.

Let $\mathcal{B}_1 = \mathcal{B}_1(\mathbf{Z}, \Omega_0, (\bar{\Omega}_\Lambda)_{\Lambda \in \mathcal{F}})$ be the Banach space of translation invariant interactions with the norm

$$\|\Phi\|_1 = |\Phi| + |\Phi|_1, \tag{5.1}$$

where

$$|\Phi|_1 = \sum_{X \ni 0} \frac{\operatorname{diam} X}{|X|} \sup_{\xi \in \Omega_X} |\Phi(\xi)| < +\infty \tag{5.2}$$

and diam $\{x_1, \ldots, x_l\} = x_l - x_1$ if $x_1 < \cdots < x_l$. Notice that $\mathcal{B}_1 \subset \mathcal{B}$ and $\| \ \|_1 \geqslant \| \ \|$ (see also Remark 5.17 below).

Let $R \subset \{x \in \mathbf{Z} : x < a\}$, $S \subset \{x \in \mathbf{Z} : x \geqslant a\}$. If $\Phi \in \mathcal{B}_1, W_{RS}$ is well defined even for infinite R, S, and we have

$$|W_{R,S}(\xi)| \leqslant \sum_{\substack{X : X \cap R \neq \emptyset, \\ X \cap S \neq \emptyset}} |\Phi(\xi|X)| \leqslant |\Phi|_1. \tag{5.3}$$

5.1 Lemma

If $F : (\mathbf{Z}, \Omega'_0, (\bar{\Omega}'_{\Lambda'})_{\Lambda' \in \mathcal{F}})| \to (\mathbf{Z}, \Omega_0, (\bar{\Omega}_\Lambda)_{\Lambda \in \mathcal{F}})$ *is a* $\mathbf{Z}$*-morphism, and* $\Phi \in \mathcal{B}_1(\mathbf{Z}, \Omega_0, (\bar{\Omega}_\Lambda)_{\Lambda \in \mathcal{F}})$*, then* $F^*\Phi \in \mathcal{B}_1(\mathbf{Z}, \Omega'_0, (\bar{\Omega}'_{\Lambda'})_{\Lambda' \in \mathcal{F}})$ *and the map* F^* *is continuous.*

From section 2.3 we find indeed

$$|F^*\Phi| \leqslant |M\ (0)| \cdot |\Phi|,$$

and similarly

$$|F^*\Phi|_1 \leqslant |M(0)| \cdot [(\operatorname{diam} M(0)) \cdot |\Phi| + |\Phi|_1].$$

Theorems 5.2 and 5.3 below show that the study of Gibbs states on a $\mathbf{Z}$-lattice system, with an interaction $\Phi \in \mathcal{B}_1$, reduces to the study of Gibbs states on mixing $\mathbf{Z}$-lattice systems.

5.2 Theorem

Given a $\mathbf{Z}$*-lattice system* $(\mathbf{Z}, \Omega_0, (\bar{\Omega}_\Lambda)_{\Lambda \in \mathcal{F}})$*, there exist finitely many transitive* $(\Omega_0^{(\alpha)}, t^{(\alpha)})$*, and injective* $\mathbf{Z}$*-morphisms*

$$F^{(\alpha)} : (\Omega_0^{(\alpha)}, t^{(\alpha)}) \mapsto (\mathbf{Z}, \Omega_0, (\bar{\Omega}_\Lambda)_{\Lambda \in \mathcal{F}}),$$

such that

(a) *the images* $F^{(\alpha)}\Omega^{(\alpha)}$ *are disjoint;*
(b) *every Gibbs state for an interaction* $\Phi \in \mathcal{B}_1$ *is a convex combination of Gibbs states* $F^{(\alpha)}\sigma^{(\alpha)}$ *where* $\sigma^{(\alpha)}$ *is a Gibbs state for* $F^{(\alpha)*}\Phi$ *on* $(\Omega_0^{(\alpha)}, t^{(\alpha)})$.
(c) *If* $\xi \in \Omega$ *is periodic (i.e.,* $\tau^p\xi = \xi$ *with* $p > 0$*), then* ξ *is in one of the* $F^{(\alpha)}\Omega^{(\alpha)}$.

5.3 Theorem

Given a transitive $\mathbf{Z}$*-lattice system* $(\mathbf{Z}, \Omega_0, (\bar{\Omega}_\Lambda)_{\Lambda\in\mathcal{F}})$*, there exists an integer* $N > 0$*,* N *mixing* $N\mathbf{Z}$*-lattice systems** $(\Omega_0^{(\beta)}, t^{(\beta)})$*, and injective* $N\mathbf{Z}$ *morphisms*

$$F^{(\beta)} : \left(\Omega_0^{(\beta)}, t^{(\beta)}\right) \to \left(N\mathbf{Z}, \Omega_{M(0)}, (\bar{\Omega}'_{\Lambda'})_{\Lambda'\in\mathcal{F}'}\right),$$

where $(N\mathbf{Z}, \Omega_{M(0)}, (\bar{\Omega}'_{\Lambda'})_{\Lambda'\in\mathcal{F}'})$ *is obtained from* $(\mathbf{Z}, \Omega_0, (\bar{\Omega}_\Lambda)_{\Lambda\in\mathcal{F}})$ *by restriction of* $\mathbf{Z}$ *to the subgroup* $N\mathbf{Z}$*, with the following properties.*

(a) *The images* $F^{(\beta)}\Omega^{(\beta)}$ *form a partition of* Ω *and are cyclically permuted by* τ.
(b) *Let* $\Phi \in \mathcal{B}_1(\mathbf{Z}, \Omega_0, (\bar{\Omega}_\Lambda)_{\Lambda\in\mathcal{F}})$ *and* Φ^* *be the corresponding interaction for* $(N\mathbf{Z}, \Omega_{M(0)}, (\bar{\Omega}'_{\Lambda'})_{\Lambda'\in\mathcal{F}'})$ *defined in Proposition 4.14. Then every Gibbs state for* Φ *is a convex combination of Gibbs states* $F^{(\beta)}\sigma^{(\beta)}$ *where* $\sigma^{(\beta)}$ *is the unique Gibbs state for* $F^{(\beta)*}\Phi^*$ *on* $(\Omega_0^{(\beta)}, t^{(\beta)})$.
(c) *If* $\xi \in \Omega$ *is periodic and* $\tau^p\xi = \xi$*, then* p *is a multiple of* N.

To prove the above theorems we shall assume, as we may, that $(\mathbf{Z}, \Omega_0, (\bar{\Omega}_\Lambda)_{\Lambda\in\mathcal{F}}) = (\Omega_0, t)$. There exists an integer $J > 0$ such that $(t^{2J})_{ij} > 0$ if and only if $(t^J)_{ij} > 0$. [Clearly one can find integers $k, l > 0$ such that $l \geqslant 2k$ and $(t^k)_{ij} > 0$ if and only if $(t^l)_{ij} > 0$. Take $J = l^2 - kl$ and notice that $J = lk + (l - 2k)l, 2J = ll + (l - 2k)l$.]

Define

$$(i \prec j) \Leftrightarrow (t^J)_{ij} > 0.$$

The relation $\prec$ is independent of the choice of J [because $i \prec j$ is equivalent to $(t^{JJ'})_{ij} > 0$ for every integer $J' > 0$]. Furthermore $i \prec j$ and $j \prec k$ imply $i \prec k$. On the set $\{i \in \Omega_0 : i \prec i\}$ we define the equivalence relation $\sim$ by

$$(i \sim j) \Leftrightarrow (i \prec j \text{ and } j \prec i).$$

* Using the isomorphism $\mathbf{Z} \mapsto N\mathbf{Z}$, we consider the $\mathbf{Z}$-lattice system $(\Omega_0^{(\beta)}, t^{(\beta)})$ as a $N\mathbf{Z}$-lattice system.

Let $[i]$ be the equivalence class of i for this relation. An order $\prec$ among equivalence classes is defined by*

$$([i] \prec [j]) \Leftrightarrow (i \prec j).$$

5.4 Lemma

Let $a, b \in \mathbf{Z}$, $a < b$, *and*

$$\Lambda = \{x \in \mathbf{Z} : aJ < x < bJ\}, \quad M = \{x \in \mathbf{Z} : (a-1)J < x < (b+1)J\}.$$

There exists $K > 0$, *independent of* a, b, *such that if* $\eta, \eta' \in \Omega_{\mathbf{Z}\setminus M}$ *and*

$$[\eta_{(a-1)J}] = [\eta'_{(a-1)J}] \prec [\eta_{(b+1)J}] = [\eta'_{(b+1)J}],$$

then

$$(\alpha_{\Lambda M}\mu_{(M)\eta})\{\xi\} \leqslant K\ (\alpha_{\Lambda M}\mu_{(M)\eta'})\{\xi\}. \tag{5.4}$$

First we notice that if $\zeta, \zeta' \in \Omega$ and $\zeta|\Lambda = \zeta'|\Lambda$, then

$$\exp[-U(\zeta|M) - W_{M,\mathbf{Z}\setminus M}(\zeta)] \leqslant K_1 \exp[-U(\zeta'|M) - W_{M,\mathbf{Z}\setminus M}(\zeta')],$$

where $K_1 = \exp[2 \times 2 \times (2|\Phi|_1 + J|\Phi|)]$. Therefore

$$\begin{aligned} &\sum_{\substack{\zeta^*:\zeta^*|\Lambda=\zeta,\\ \zeta^*|(\mathbf{Z}\setminus M)=\eta}} \exp[-U(\zeta^*|M) - W_{M,\mathbf{Z}\setminus M}(\zeta^*)] \\ &\leqslant K_2 \sum_{\substack{\zeta^*:\zeta^*|\Lambda=\zeta,\\ \zeta^*|(\mathbf{Z}\setminus M)=\eta'}} \exp[-U(\zeta^*|M) - W_{M,\mathbf{Z}\setminus M}(\zeta^*)], \end{aligned} \tag{5.5}$$

where $K_2 = |\Omega_0|^{2J} K_1$. Summing over ζ and interchanging η, η' yields

$$\begin{aligned} &\sum_{\zeta^*:\zeta^*|(\mathbf{Z}\setminus M)=\eta} \exp[-U(\zeta^*|M) - W_{M,\mathbf{Z}\setminus M}(\zeta^*)] \\ &\geqslant (K_2)^{-1} \sum_{\zeta^*:\zeta^*|(\mathbf{Z}\setminus M)=\eta'} \exp[-U(\zeta^*|M) - W_{M,\mathbf{Z}\setminus M}(\zeta^*)]. \end{aligned} \tag{5.6}$$

Dividing (5.5) by (5.6) yields (5.4) with $K = (K_2)^2$.

* The definition of equivalence classes and order in this context is standard in the theory of Markov chains. See for instance Chung [1], I, Section 3.

5.5 Proof of theorems 5.2 and 5.3

Given classes $[i]$, $[j]$, define a function C on Ω by

$$C(\xi) = \begin{cases} 1 & \text{if } \lim_{n\to-\infty} [\xi_{nJ}] = [i] \quad \text{and} \quad \lim_{n\to+\infty} [\xi_{nJ}] = [j], \\ 0 & \text{otherwise.} \end{cases}$$

For any probability measure σ on Ω, C belongs to the algebra at infinity (see Section 1.10), and therefore if σ is a pure (= extremal) Gibbs state, C is almost everywhere 0 or 1. Therefore, given a pure Gibbs state σ for an interaction $\Phi \in \mathcal{B}$, there are (uniquely defined) classes $[i]$, $[j]$ such that

$$\lim_{n\to-\infty} [\xi_{nJ}] = [i], \qquad \lim_{n\to+\infty} [\xi_{nJ}] = [j]$$

σ-almost everywhere.

Suppose first that $[i] \neq [j]$. Given $\varepsilon > 0$, there exists N such that

$$\sigma(\{\xi : [\xi_{nJ}] = [i] \text{ for } n \leqslant -N \text{ and } [\xi_{nJ}] = [j] \text{ for } n \geqslant N\}) > 1 - \varepsilon.$$

Since σ is a Gibbs state, we can estimate the probability p_- that $[\xi_{-NJ}] = [i]$ in terms of $\mu_{(M)\eta}\{\xi\}$, where M is a large interval centered at $-NJ$ and

$$[\eta_{nJ}] = \begin{cases} [i] & \text{for} \quad n < 0, \\ [j] & \text{for} \quad n > 0. \end{cases}$$

Similarly for the probability p_+ that $[\xi_{nJ}] = [i]$. By translation invariance of Φ and Lemma 5.4 we have

$$p_- \leqslant K p_+.$$

This is incompatible with $p_- > 1 - \varepsilon$, $p_+ < \varepsilon$ if ε has been chosen sufficiently small. Therefore we cannot have $[i] \neq [j]$.

Suppose now that $[i] = [j]$. We show that there is only one pure Gibbs state in this situation. Let Λ, M be as in Lemma 5.4, and $[\eta_{nJ}] = [\eta'_{nJ}] = [i]$ for all n. Fixing η', we define

$$\tilde{\mu}_{(\Lambda)\eta}\{\xi\} = (\alpha_{\Lambda M}\mu_{(M)\eta})\{\xi\} - \frac{1}{K}(\alpha_{\Lambda M}\mu_{(M)\eta'})\{\xi\} \geqslant 0.$$

Then

$$\|\tilde{\mu}_{(\Lambda)\eta}\| = \sum_{\xi\in\Omega_\Lambda} \tilde{\mu}_{(\Lambda)_\eta}\{\xi\} = 1 - K^{-1}.$$

If σ, σ' are two Gibbs states, we have

$$\alpha_\Lambda(\sigma - \sigma')\{\xi\} = \int [\sigma_{\mathbf{Z}\setminus M}(d\eta) - \sigma'_{\mathbf{Z}\setminus M}(d\eta)]\tilde{\mu}_{(\Lambda)_\eta}\{\xi\};$$

hence

$$\|\sigma - \sigma'\| = \lim_{-a,b\to\infty} \|\alpha_\Lambda(\sigma - \sigma')\| \leqslant \|\sigma - \sigma'\|(1 - K^{-1}),$$

so that $\|\sigma - \sigma'\| = 0$.

We have thus shown that for each pure Gibbs state σ on (Ω_0, t) there is a different class $[i]$ such that σ is carried by

$$\Omega^{[i]} = \{\xi \in \Omega : [\xi_{nJ}] = [i] \text{ for all } n \in \mathbf{Z}\}.$$

Let $\xi, \xi' \in \Omega^{[i]}$ and $0 < a < J$; then $(t^a)_{i\xi_a} > 0$ and $(t^{J-a})_{\xi'_a i'} > 0$ for some $i, i' \in [i]$. Since $(t^J)_{i,i'} > 0$, we have $(t^{2J})_{\xi'_a \xi_a} > 0$; hence $\xi'_a \prec \xi_a$. Therefore ξ_a and ξ'_a are in the same class $[j]$. This shows that $\tau^a\Omega^{[i]} \subset \Omega^{[j]}$. Therefore similarly $\tau^{J-a}\Omega^{[j]} \subset \Omega^{[i]}$, so that $\tau^a\Omega^{[i]} = \Omega^{[j]}$.

Let α be the set of classes $[j]$ such that $\Omega^{[j]} = \tau^a\Omega^{[i]}$ for some $a \in \mathbf{Z}$ (or equivalently $0 \leqslant a < J$). If $[j], [k] \in \alpha$ and $[j] \prec [k]$, we show that $[j] = [k]$. We have $\tau^a\Omega^{[k]} = \Omega^{[j]}$ for some a, $0 \leqslant a < J$; hence $(t^a)_{kj} > 0$ for some $j \in [j], k \in [k]$. On the other hand $[j] \prec [k]$ means $(t^J)_{jk} > 0$; therefore $(t^{(l+1)a+lJ})_{kj} > 0$ for all integers $l \geqslant 0$. Taking $l + 1 = J$ gives $k \prec j$, so that $[j] = [k]$. We have thus proved that *distinct classes* $[j], [k] \in \alpha$ *are not comparable*.

Let $\Omega_0^{(\alpha)} \subset \Omega_0$ be the union of the classes $[j] \in \alpha$, let $t^{(\alpha)}$ be the restriction of t to $\Omega_0^{(\alpha)} \times \Omega_0^{(\alpha)}$, and let

$$\Omega^{(\alpha)} = \left\{\xi \in \left(\Omega_0^{(\alpha)}\right)^{\mathbf{Z}} : t^{(\alpha)}_{\xi_x\xi_{x+1}} = 1 \text{ for all } x \in \mathbf{Z}\right\}.$$

We define $F^{(\alpha)} : \Omega^{(\alpha)} \mapsto \Omega$ to be the inclusion map. If $\xi \in F^{(\alpha)}\Omega^{(\alpha)}$, let $\lim_{n\to-\infty}[\xi_{nJ}] = [j]$ and $\lim_{n\to+\infty}[\xi_{nJ}] = [k]$. Then $[j] \prec [k]$; hence $[j] = [k]$ and $\xi \in \Omega^{[j]}$. Therefore

$$F^{(\alpha)}\Omega^{(\alpha)} = \bigcup_{[j]\in\alpha} \Omega^{[j]}. \tag{5.7}$$

We can now prove Theorem 5.2. First it is clear that for all $j, k \in \Omega_0^{(\alpha)}$ there is $a > 0$ such that $(t^a)_{jk} > 0$, and therefore $(\Omega_0^{(\alpha)}, t^{(\alpha)})$ is transitive. It is easy to see that $F^{(\alpha)}$ is a $\mathbf{Z}$-morphism; the main point is to check that if $\xi \in \Omega^{(\alpha)}$ then $\sum_{F^{(\alpha)}\xi} \subset F^{(\alpha)}\Omega^{(\alpha)}$. Since we may (by (5.7)) suppose $F^{(\alpha)}\xi \in \Omega^{[j]}$, we have indeed $\sum_{F^{(\alpha)}\xi} \subset \Omega^{[i]} \subset F^{(\alpha)}\Omega^{(\alpha)}$. The distinct sets α of classes $[i]$ which can be defined are disjoint, proving part (a) of Theorem 5.2. Remember now that every pure Gibbs state of $\Phi \in \mathcal{B}_1$ on Ω has its support in one of the $\Omega^{[i]}$ and therefore in one of the $F^{(\alpha)}\Omega^{(\alpha)}$. Such a state, restricted to $\Omega^{(\alpha)}$, is then also a Gibbs state for the restriction of Φ to $\cup\,\{\Omega_X^{(\alpha)} : X \text{ finite}\}$, i.e. for $F^{(\alpha)*}\Phi$. This proves (b). The proof of (c) is immediate.

To prove Theorem 5.3, we assume that (Ω_0, t) is transitive, i.e., there is only one α and $\Omega = \Omega^{(\alpha)}$. Let N be the number of distinct classes $[j]$ in α, then the $\Omega^{[j]}$ are cyclically permuted by τ with period N, and (5.7) becomes

$$\Omega = \bigcup_{\beta=0}^{N-1} \tau^{-\beta}\Omega^{[i]}. \tag{5.8}$$

The map $x \mapsto Nx$ is an isomorphism of $\mathbf{Z}$ on its subgroup $N\mathbf{Z}$. Restriction of $\mathbf{Z}$ to $N\mathbf{Z}$ in (Ω_0, t) yields a $N\mathbf{Z}$-lattice system which (in view of the isomorphism $\mathbf{Z} \mapsto N\mathbf{Z}$) may be considered as a $\mathbf{Z}$-lattice system (Ω_0^*, t^*). Here we take $\Omega_0^* = \Omega_{[0,N)}$, and if $\xi, \xi' \in \Omega_0^*$, $\xi = (\xi_0, \xi_1, \ldots, \xi_{N-1})$, $\xi' = (\xi'_0, \xi'_1, \ldots, \xi'_{N-1})$, we write $t^*_{\xi\xi'} = t_{\xi_{N-1}\xi'_0}$. For $\beta = 0, 1, \ldots, N-1$, let $\Omega_0^{(\beta)} = \{(\xi_0, \xi_1, \ldots, \xi_{N-1}) \in \Omega_0^* : \xi_\beta \in [i]\}$. If ξ, ξ' belong to different $\Omega_0^{(\beta)}$, then $t^*_{\xi\xi'} = 0$. We let $t^{(\beta)}$ be the restriction of t^* to $\Omega_0^{(\beta)} \times \Omega_0^{(\beta)}$. Define

$$\Omega^{(\beta)} = \left\{\xi \in \left(\Omega_0^{(\beta)}\right)^{\mathbf{Z}} : t^{(\beta)}_{\xi_x\xi_{x+1}} = 1 \text{ for all } x \in \mathbf{Z}\right\},$$

and $F^{(\beta)} : \Omega^{(\beta)} \mapsto \Omega$ be the inclusion; then

$$F^{(\beta)}\Omega^{(\beta)} = \tau^{-\beta}\Omega^{[i]} = \Omega^{[j]}$$

for some $[j]$. Since N is a divisor of J, $(\Omega_0^{(\beta)}, t^{(\beta)})$ is mixing by definition of the equivalence classes $[j]$. Part (a) of Theorem 5.3 follows then from (5.8). To prove (b) we notice that if $\Phi \in \mathcal{B}_1(\Omega_0, t)$, then the corresponding interaction Φ^* on (Ω^*, t^*) is in $\mathcal{B}_1(\Omega_0^*, t^*)$. It suffices then to apply Theorem 5.2(b) and the fact proved above that the pure Gibbs states are carried by the $\Omega^{[j]}$. The proof of (c) is immediate.

5.6 Corollaries to theorems 5.2 and 5.3

Let $(\mathbf{Z}, \Omega_0, (\bar{\Omega}_\Lambda)_{\Lambda\in\mathcal{F}})$ *be a* $\mathbf{Z}$*-lattice system and* Φ *an interaction in* $\mathcal{B}_1$.

(a) *If* $(\mathbf{Z}, \Omega_0, (\bar{\Omega}_\Lambda)_{\Lambda\in\mathcal{F}})$ *is transitive, there is a unique equilibrium state,* which coincides with the unique translation invariant Gibbs state; its support is* Ω.
(b) *If* $(\mathbf{Z}, \Omega_0, (\bar{\Omega}_\Lambda)_{\Lambda\in\mathcal{F}})$ *is mixing, there is a unique Gibbs state.*
(c) *In the situation of Theorem 5.2, we have* (*with* $A \in \mathcal{C}$)

$$P^\Phi = \max_\alpha P^{F^{(\alpha)}*\Phi}, \qquad P(A) = \max_\alpha P(A \circ F^{(\alpha)}).$$

* One can prove that there is a unique equilibrium state for A under slightly less restrictive conditions than $A = A_\Phi$ with $\Phi \in \mathcal{B}_1$. See section 7.14.

(d) *In the situation of Theorem 5.3, we have for all* β

$$P^{\Phi} = N^{-1} P^{F^{(\beta)*}\Phi^*}, \quad P(A) = N^{-1} P\big((A + A \circ \tau + \cdots + A \circ \tau^{N-1}) \circ F^{(\beta)}\big).$$

In the transitive case, an invariant Gibbs state is necessarily of the form

$$\frac{1}{N} \sum_{\beta=0}^{N-1} \tau^{-\beta} \sigma,$$

where σ is a Gibbs state with support in $\Omega^{[i]}$ and therefore unique. By remark 1.14, supp $\sigma = \Omega^{[i]}$. There is thus at most one invariant Gibbs state, and its support is Ω. Since there is at least one equilibrium state, Theorem 4.2 proves (a); (b) is immediate.

(c) follows from the variational principle for P (Theorem 3.12) and Proposition 4.12(a).

(d) follows from (c) and Proposition 4.14(b).

5.7 Theorem

Let $\Phi, \Phi' \in \mathcal{B}_1$ *for a transitive* **Z**-*lattice system, and let* ρ, ρ' *be the corresponding equilibrium states. Then* $\rho = \rho'$ *if and only if there exist* $c \in \mathbf{R}$ *and* $C \in \mathcal{C}$ *such that*

$$A_{\Phi'} - A_{\Phi} = c + C \circ \tau - C. \tag{5.9}$$

In this formula c *is uniquely determined by* $\Phi' - \Phi$, *and* C *is determined up to an additive constant.*

If (5.9) holds, then $P(A_{\Phi'} + B) = P(A_{\Phi} + B) + c$ for all $B \in \mathcal{C}$ (Theorem 3.4), and it follows that $\rho = \rho'$ (Theorem 3.7).

Conversely, let $\rho = \rho'$. By convexity of P, ρ is an equilibrium state for $(1 - t)\Phi + t\Phi'$ whenever $t \in [0, 1]$. We are thus in the situation discussed in Section 4.6: ρ is a Gibbs state for $(1 - t)\Phi + t\Phi'$ when $t \in [0, 1]$ and, by analytic continuation, for all real t. Since ρ is a translation invariant Gibbs state, and the **Z**-lattice system is transitive, Corollary 5.6(a) shows that ρ is an equilibrium state for $(1 - t)\Phi + t\Phi'$ (all real t). Let $\Psi = \Phi' - \Phi$ and $c = \rho(A_{\Psi})$; then

$$\max_{\sigma \in I}[s(\sigma) + \sigma(A_{\Phi}) + t\sigma(A_{\Psi})] = s(\rho) + \rho(A_{\Phi + t\Psi}) = P^{\Phi} + ct$$

for all $t \in \mathbf{R}$. This implies $\sigma(A_{\Psi}) = c$ for all $\sigma \in I$, or

$$A_{\Phi'} - A_{\Phi} = c + B, \tag{5.10}$$

where $\sigma(B) = 0$ for all $\sigma \in I$. In particular, if $\xi \in \Omega$ has period n,

$$\sum_{j=0}^{n-1} B(\tau^j \xi) = 0. \tag{5.11}$$

Since the **Z**-lattice system is transitive, there is $\eta \in \Omega$ with dense orbit $\Gamma = \{\tau^k \eta : k \in \mathbf{Z}\}$. Define C on Γ by

$$C(\tau^k \eta) = \begin{cases} \sum_{j=0}^{k-1} B(\tau^j \eta) & \text{if} \quad k \geqslant 0, \\ -\sum_{j=k}^{-1} B(\tau^j \eta) & \text{if} \quad k < 0. \end{cases}$$

Let $\eta', \eta'' \in \Gamma$ and $\eta'_x = \eta''_x$ for $|x| \leqslant m$. If $\eta' = \tau^k \eta$, $\eta'' = \tau^l \eta$, and $k < l$, then

$$C(\eta'') - C(\eta') = \sum_{j=k}^{l-1} B(\tau^j \eta)$$

and $\eta_{k+x} = \eta_{l+x}$ for $|x| \leqslant m$. If m is large enough, then there exists $\xi \in \Omega$ such that $\tau^{l-k}\xi = \xi$ and $\xi_x = \eta_x$ when $k - m \leqslant x \leqslant l + m$. Using (5.11),

$$C(\eta'') - C(\eta') = \sum_{j=k}^{l-1} [B(\tau^j \eta) - B(\tau^j \xi)].$$

If $(\tau^j \eta)_x = (\tau^j \xi)_x$ for $|x| \leqslant r$, we have

$$\begin{aligned} |B(\tau^j \eta) - B(\tau^j \xi)| &= |A_\Psi(\tau^j \eta) - A_\Psi(\tau^j \xi)| \\ &\leqslant 2 \sum_{X \ni 0 \,:\, \operatorname{diam} X > r} \frac{1}{|X|} \sup_{\xi \in \Omega_X} |\Psi(\xi)|. \end{aligned} \tag{5.12}$$

Therefore

$$\begin{aligned} |C(\eta'') - C(\eta')| &\leqslant 4 \sum_{r=m}^{\infty} \sum_{X \ni 0 \,:\, \operatorname{diam} X > r} \frac{1}{|X|} \sup_{\xi \in \Omega_X} |\Psi(\xi)| \\ &\leqslant 4 \sum_{X \ni 0 \,:\, \operatorname{diam} X > m} \frac{\operatorname{diam} X}{|X|} \sup_{\xi \in \Omega_X} |\Psi(\xi)|. \end{aligned}$$

Since $\Psi \in \mathcal{B}_1$, this is small when m is large, showing that C extends to a continuous function on Ω. Clearly

$$C \circ \tau - C = B$$

and therefore (5.9) follows from (5.10).

It is clear from the construction that c is unique, and C unique up to an additive constant.

5.8 Mixing Z-lattice systems

In view of Theorems 5.2 and 5.3 we shall now concentrate our attention on mixing $\mathbf{Z}$-lattice systems. Thus let (Ω_0, t) be mixing. We use the notation $\mathbf{Z}_\geqslant, \mathbf{Z}_\leqslant, \mathbf{Z}_>, \mathbf{Z}_<$ for the sets of integers which are respectively $\geqslant 0$, $\leqslant 0$, > 0, < 0. We define then

$$\Omega_\geqslant = \{\xi \in (\Omega_0)^{\mathbf{Z}_\geqslant} : t_{\xi_x \xi_{x+1}} = 1 \text{ for all } x \in \mathbf{Z}_\geqslant\},$$

and similarly for $\Omega_\leqslant, \Omega_>, \Omega_<$. Given an interaction $\Phi \in \mathcal{B}_1(\Omega_0, t)$, we can restrict it to the subsets of $\mathbf{Z}_\geqslant, \mathbf{Z}_\leqslant, \mathbf{Z}_>, \mathbf{Z}_<$, obtaining interactions $\Phi_\geqslant, \Phi_\leqslant, \Phi_>, \Phi_<$.

5.9 Lemma

(a) *There is a unique Gibbs state $\sigma_\geqslant$ (respectively $\sigma_\leqslant, \sigma_>, \sigma_<$) on $\Omega_\geqslant$ (respectively $\Omega_\leqslant, \Omega_>, \Omega_<$) for the interaction $\Phi_\geqslant$ (respectively $\Phi_\leqslant, \Phi_>, \Phi_<$).*

(b) *There is $K > 0$ such that*

$$\sigma_\leqslant(d\xi_\leqslant) \times \sigma_>(d\xi_>) \exp[-W_{\mathbf{Z}_\leqslant, \mathbf{Z}_>}(\xi_\leqslant \vee \xi_>)] = K\sigma(d(\xi_\leqslant \vee \xi_>)),$$

where σ is the unique Gibbs state for Φ on Ω.

(c) *Given $\varepsilon > 0$ there exists $n(\varepsilon)$ such that if $B \in \mathcal{C}(\Omega_\geqslant \times \Omega_>)$, $\|B\| \leqslant 1$, and $B(\xi_\leqslant, \xi_>)$ does not depend on ξ_x for $|x| < n(\varepsilon)$, then*

$$\left| \int \sigma(d(\xi_\leqslant \vee \xi_<)) B(\sigma_\leqslant, \xi_>) \right. \\ \left. - \int\int \sigma(d(\xi'_\leqslant \vee \xi'_>)) \sigma(d(\xi''_\leqslant \vee \xi''_>)) B(\xi'_\leqslant, \xi''_>) \right| < \varepsilon.$$

To establish (a) it suffices to follow the proof (Section 5.5 and Lemma 5.4) that there is a unique Gibbs state with support in $\Omega^{[i]}$, replacing $\mathbf{Z}$ by $\mathbf{Z}_\geqslant$ (respectively $\mathbf{Z}_\leqslant, \mathbf{Z}_>, \mathbf{Z}_<$) and $[i]$ by Ω_0.

The fact that $\Phi \in \mathcal{B}_1$ ensures that $\exp[-W_{\mathbf{Z}_\leqslant, \mathbf{Z}_>}]$ is continuous on $\Omega_\leqslant \times \Omega_>$. Using the definition of Gibbs states then yields (b).

Consider the lattice system $(L, (\Omega_x)_{x \in L}, (\bar{\Omega}_\Lambda)_{\Lambda \in \mathcal{F}})$ where $L = \mathbf{Z} = \mathbf{Z}_\leqslant \cup \mathbf{Z}_>$, $\Omega_x = \Omega_0$,

$$\mathcal{F} = \{\{x, x+1\} : x \in \mathbf{Z} \text{ and } x \neq 0\},$$
$$\bar{\Omega}_{(x,x+1)} = \{(\xi_x, \xi_{x+1}) : t_{\xi_x \xi_{x+1}} = 1\}.$$

The configuration space of this lattice system is $\Omega_\leqslant \times \Omega_>$. An interaction Φ^* is defined such that $\Phi^*(\xi|X) = \Phi(\xi|X)$ if $X \subset \mathbf{Z}_\leqslant$ or $X \subset \mathbf{Z}_>$, and $\Phi^*(\xi|X) = 0$ otherwise. Using the definition of Gibbs states and (a), it is clear that there is only one Gibbs state for Φ^*, namely $\sigma_\leqslant \otimes \sigma_>$. [For a generalization, see Exercise 4 of Chapter 2.]

We may thus apply Theorem 1.11(C) with

$$A(\xi_\leqslant, \xi_>) = \frac{2}{K\varepsilon} \exp[-W(\xi_\leqslant \vee \xi_>)],$$

where we have written $W = W_{\mathbf{Z}_\leqslant, \mathbf{Z}_>}$. Using (b) we find that if $B \in \mathcal{C}(\Omega_\leqslant \times \Omega_>)$, $\|B\| \leqslant 1$, and $B(\xi)$ does not depend on ξ_x for $|x| < n$, then

$$\left| \int \sigma(d\,(\xi_\leqslant \vee \xi_>))B(\xi_\leqslant, \xi_>) - \int\int \sigma_\leqslant(d\xi_\leqslant)\sigma_>(d\xi_>)B(\xi_\leqslant, \xi_>) \right| < \frac{\varepsilon}{2} \tag{5.13}$$

for sufficiently large n. Similarly, taking

$$A(\xi'_\leqslant, \xi''_>) = \frac{2}{K^2\varepsilon}\left[\int \sigma_>(d\xi'_>)\exp[-W\,(\xi'_\leqslant \vee \xi'_>)]\right] \times \left[\int \sigma_\leqslant(d\xi''_\leqslant)\exp[-W\,(\xi''_\leqslant \vee \xi''_>)]\right],$$

we find

$$\left| \int\int \sigma_\leqslant(d\xi_\leqslant)\sigma_>(d\xi_>)B(\xi_\leqslant, \xi_>) - \int\int \sigma(d\,(\xi'_\leqslant \vee \xi'_>))\sigma(d\,(\xi''_\leqslant \vee \xi''_>))B(\xi'_\leqslant, \xi''_>) \right| < \frac{\varepsilon}{2} \tag{5.14}$$

for sufficiently large n. Comparison of (5.13) and (5.14) yields (c).

5.10 Theorem*

Let $\Phi \in \mathcal{B}_1$ be an interaction for the mixing **Z**-*lattice system (Ω_0, t), and σ the unique Gibbs state. If Ω consists of more than one point, the dynamical system (Ω, σ, τ) is equivalent to a Bernoulli shift.*

The sets $\mathfrak{A}_i = \{\xi \in \Omega : \xi_0 = i\}$ form a partition $\mathfrak{A}$ of Ω. According to the Friedman-Ornstein theorem (see Appendix A.4.6) it suffices to show that $\mathfrak{A}$ is weak Bernoulli for the system (Ω, σ, τ). This means that, given $\varepsilon > 0$, there exists $n(\varepsilon)$ such that for $n \geqslant n(\varepsilon)$, and all $k, l \geqslant 0$,

$$\sum_{\eta,\zeta} |(\alpha_{[-n-k,-n]\cup[n,n+l]}\sigma)\{\eta \vee \zeta\} - (\alpha_{[-n-k,-n]}\sigma)\{\eta\} \cdot (\alpha_{[n,n+l]}\sigma)\{\zeta\}| < \varepsilon. \tag{5.15}$$

* See Gallavotti [1], Ledrappier [1].

Let $\hat{B}(\eta \vee \zeta)$ be ± 1 depending on the sign of the quantity between the modulus bars. If $\xi_{\leqslant}|[-n-k,-n] = \eta$ and $\xi_{>}|[n,n+l] = \zeta$, we put $B(\xi_{\leqslant},\xi_{>}) = \hat{B}(\eta \vee \zeta)$. With this choice of B, (5.15) results from Lemma 5.9(c).

5.11 The transfer matrix and the operator $\mathfrak{L}$

If $\xi_{\geqslant} \in \Omega_{\geqslant}$ we write $\xi_{\geqslant} = \xi_0 \vee \xi_{>}$ with $\xi_0 \in \Omega_0$, $\xi_{>} \in \Omega_{>}$. Given a measure μ on $\Omega_{>}$, let $\mathfrak{M}^*\mu$ be the measure on $\Omega_{\geqslant}$ such that

$$(\mathfrak{M}^*\mu)(A) = \sum_{\xi_0} \int \mu(d\xi_{>})A(\xi_0 \vee \xi_{>})$$
$$\exp[-U_{\{0\}}(\xi_0) - W_{\{0\},\mathbf{Z}_{>}}(\xi_0 \vee \xi_{>})].$$

From the definition of Gibbs states, we find that $\mathfrak{M}^*\sigma_{>}$ is proportional to a Gibbs state for $\Phi_{\geqslant}$ on $\Omega_{\geqslant}$. Therefore, by Lemma 5.9(a),

$$\mathfrak{M}^*\sigma_{>} = \lambda\sigma_{\geqslant}$$

with some $\lambda > 0$. The map $\tau^{-1} : \Omega_{\geqslant} \mapsto \Omega_{>}$ is defined according to Section 3.1, and it is clear that $\tau^{-1}\sigma_{\geqslant} = \sigma_{>}$. Therefore

$$\mathfrak{L}^*\sigma_{>} = \lambda\sigma_{>}, \tag{5.16}$$

where we have written

$$\mathfrak{L}^* = \tau^{-1}\mathfrak{M}^*.$$

The operator $\mathfrak{L}^*$ on measures is called the *transfer matrix*. It is the adjoint of an operator $\mathfrak{L}$ on $\mathcal{C}(\Omega_{>})$ defined by

$$(\mathfrak{L}A)(\xi_{>}) = \sum_{\xi_0} A(\tau^{-1}(\xi_0 \vee \xi_{>}))\exp[-U_{\{0\}}(\xi_0) - W_{\{0\},\mathbf{Z}_{>}}(\xi_0 \vee \xi_{>})].$$

Notice that

$$\mathfrak{L}[A \cdot (B \circ \alpha_{\mathbf{Z}_{>}\mathbf{Z}_{\geqslant}} \circ \tau)] = (\mathfrak{L}A) \cdot B \tag{5.17}$$

when $A, B \in \mathcal{C}(\Omega_{>})$.

As we shall see, the study of the operator $\mathfrak{L}$ yields important results on Gibbs states and the pressure function in the case of exponentially decreasing interactions.

5.12 The function $\psi_>$

A continuous function $\psi_>$ is defined on $\Omega_>$ by

$$\psi_>(\xi_>) = \frac{1}{K}\int \sigma_\leqslant(d\xi_\leqslant)\exp[-W_{\mathbf{Z}_\leqslant,\mathbf{Z}_>}(\xi_\leqslant \vee \xi_>)],$$

where K is given by Lemma 5.9(b). In view of this lemma

$$\sigma_>(\psi_> \cdot A) = \sigma(A \circ \alpha_{\mathbf{Z}_>}) \tag{5.18}$$

when $A \in \mathcal{C}(\Omega_>)$; in particular $\sigma_>(\psi_>) = 1$.

Since (Ω_0, t) is mixing and $\sigma_\leqslant$ is a Gibbs state for $\Phi_\leqslant$ (respectively, $\sigma_>$ is a Gibbs state for $\Phi_>$), we know that

$$\operatorname{supp}\sigma_\leqslant = \Omega_\leqslant, \qquad \operatorname{supp}\sigma_> = \Omega_> \tag{5.19}$$

(see Remark 1.14). From the definition we also obtain

$$d^{-1} \leqslant \psi_>(\xi_>) \leqslant d \tag{5.20}$$

for some $d > 0$.

Using (5.17), (5.18) we have

$$\begin{aligned}\sigma_>((\mathfrak{L}\psi_>)\cdot B) &= \sigma_>(\mathfrak{L}[\psi_> \cdot (B \circ \alpha_{\mathbf{Z}_>\mathbf{Z}_\geqslant} \circ \tau)]) \\ &= \lambda\sigma_>(\psi_> \cdot (B \circ \alpha_{\mathbf{Z}_>\mathbf{Z}_\geqslant} \circ \tau)) = \lambda\sigma(B \circ \alpha_{\mathbf{Z}_>}) = \sigma_>((\lambda\psi_>)\cdot B)\end{aligned}$$

and therefore, using (5.19),

$$\mathfrak{L}\psi_> = \lambda\psi_>. \tag{5.21}$$

5.13 Proposition

The eigenvalue λ of $\mathfrak{L}$ and $\mathfrak{L}^*$ defined by (5.21) *and* (5.16) *satisfies*

$$\log\lambda = P^\Phi.$$

Using (5.20) we have indeed

$$\begin{aligned}\log\lambda &= \lim_{n\to\infty}\frac{1}{n}\log \mathfrak{L}^n\psi_>(\xi_>) \\ &= \lim_{n\to\infty}\frac{1}{n}\log(\mathfrak{L}^n 1)(\xi_>) \\ &= \lim_{n\to\infty}\frac{1}{n}\log\sum_{\xi_{-n+1}\ldots\xi_0}\exp[-U_{\{-n+1,\ldots,0\}}(\xi_{-n+1}\vee\cdots\vee\xi_0) \\ &\quad -W_{\{-n+1,\ldots,0\}\mathbf{Z}_>}(\xi_{-n+1}\vee\cdots\vee\xi_0\vee\xi_>)] \\ &= P^\Phi\end{aligned}$$

by Theorem 3.4.

5.14 The operator $\mathbb{S}$

In view of (5.20) a bounded operator $\mathbb{S}$ on $\mathcal{C}(\Omega_>)$ is defined by

$$\mathbb{S}A = (\psi_>)^{-1} \cdot \lambda^{-1}\mathfrak{L}(\psi_> \cdot A).$$

It satisfies

$$\mathbb{S}1 = 1,$$
$$(A \geqslant 0) \Rightarrow (\mathbb{S}A \geqslant 0);$$

and therefore

$$\|\mathbb{S}\| = 1.$$

Notice also that, by (5.17),

$$\mathbb{S}^n[A \cdot (B \circ (\alpha_{\mathbf{Z}_>\mathbf{Z}_\geqslant} \circ \tau)^n)] = (\mathbb{S}^n A) \cdot B \tag{5.22}$$

if $B \in \mathcal{C}(\Omega_>)$.

5.15 Lemma

The functions

$$\xi_> \mapsto \frac{1}{K \cdot \psi_>(\xi_>)} \int \sigma_\leqslant(d\xi_\leqslant) A(\xi_\leqslant) \exp[-W_{\mathbf{Z}_\leqslant,\mathbf{Z}_>}(\xi_\leqslant \vee \xi_>)]$$

with $A \in \mathcal{C}(\Omega_\leqslant)$ and $\|A\| \leqslant 1$, are uniformly bounded and equicontinuous on $\Omega_>$.

Let Δ be the set of these functions. We have $\mathbb{S}^n B \in \Delta$ whenever $B \in \mathcal{C}(\Omega_>)$, $\|B\| \leqslant 1$, and $B(\xi_>)$ depends only on $\xi_1, \ldots, \xi_n$.

The map $\Omega_> \mapsto \mathcal{C}(\Omega_\leqslant)$ defined by

$$\xi_> \mapsto \exp[-W_{\mathbf{Z}_\leqslant,\mathbf{Z}_>}(\cdot \vee \xi_>)]$$

is continuous. From this the boundedness and equicontinuity follow. That $\mathbb{S}^n$B $\in \Delta$ is readily checked by calculation.

5.16 Proposition

If $B \in \mathcal{C}(\Omega_>)$, then

$$\lim_{n\to\infty} \mathbb{S}^n B = \sigma(B \circ \alpha_{\mathbf{Z}_>}), \tag{5.23}$$

$$\lim_{n\to\infty} \lambda^{-n}\mathfrak{L}^n B = \sigma_>(B) \cdot \psi_>. \tag{5.24}$$

The convergence is uniform on compact subsets of $\mathcal{C}(\Omega_>)$, in particular on Δ.

Since $\|\mathbb{S}\| = 1$, it suffices to prove (5.23) under the assumption that $B(\xi_>)$ depends only on $\xi_1, \ldots, \xi_N$ for some integer N. By Lemma 5.15, the sequence $(\mathbb{S}^n B)$ has then a uniformly convergent subsequence; let $\bar{B}$ be its limit. For any $C \in \mathcal{C}(\Omega_>)$ we have, using (5.22) and (5.18),

$$\begin{aligned}
\sigma_>(\psi_> \cdot \bar{B} \cdot C) &= \lim \sigma_>(\psi_> \cdot \mathbb{S}^n B \cdot C) \\
&= \lim \sigma_>(\psi_> \cdot \mathbb{S}^n [B \cdot (C \circ (\alpha_{\mathbf{Z}_>\mathbf{Z}_\geqslant} \circ \tau)^n)]) \\
&= \lim \sigma_>(\psi_> \cdot B \cdot (C \circ (\alpha_{\mathbf{Z}_>\mathbf{Z}_\geqslant} \circ \tau)^n)) \\
&= \lim \sigma((B \circ \alpha_{\mathbf{Z}_>}) \cdot (C \circ \alpha_{\mathbf{Z}_>} \circ \tau^n)) \\
&= \sigma(B \circ \alpha_{\mathbf{Z}_>}) \cdot \sigma(C \circ \alpha_{\mathbf{Z}_>}) \\
&= \sigma(B \circ \alpha_{\mathbf{Z}_>}) \cdot \sigma_>(\psi_> \cdot C)
\end{aligned}$$

because σ is a pure Gibbs state (Theorem 1.11(C)). From (5.19), (5.20) it follows then that $\bar{B}$ is a constant: $\bar{B} = \sigma(B \circ \alpha_{\mathbf{Z}_>})$, proving (5.23). We obtain (5.24) immediately from (5.23).

Since $\|\mathbb{S}^n\| = 1$ for all $n > 0$, the convergence in (5.23) and (5.24) is uniform on compact sets.

5.17 Remark

A set $X \subset \mathbf{Z}$ is a (finite) *interval* if it is of the form $X = [k, l] = \{x : k \leqslant x \leqslant l\}$. If $\Phi \in \mathcal{B}_1$, there is an interaction $\Phi_* \in \mathcal{B}_1$ such that

(a) $\Phi_*(\xi|X) = 0$ unless X is an interval,
(b) $U_\Lambda^\Phi = U_\Lambda^{\Phi*}$ whenever Λ is an interval,
(c) $W_{\Lambda,M}^\Phi = W_{\Lambda,M}^{\Phi*}$ whenever Λ, M are intervals,
(d) $\|\Phi_*\| = \|\Phi_*\|_1 \leqslant \|\Phi\|_1$.

To see this, let Φ_* satisfy (a), and $\Phi_*(\xi|[k,l])$ be the sum of the quantities $\Phi(\xi|X)$ over all X which have k as smallest element and l as largest element; (b) and (c) are clear, and (d) follows from (5.1), (5.2).

In view of (b), (c), (d), it would have been sufficient for the purposes of this chapter to use instead of $\mathcal{B}_1$ the subspace $\mathcal{B}_*$ of $\mathcal{B}$ consisting of the interactions satisfying (a).

5.18 Exponentially decreasing interactions

If $0 < \theta < 1$, let $\mathcal{B}^\theta = \mathcal{B}^\theta(\mathbf{Z}, \Omega_0, (\bar{\Omega}_\Lambda)_{\Lambda \in \mathcal{F}})$ be the Banach space of translation invariant interactions Φ such that $\Phi(\xi|X) = 0$ unless X is an interval, with the norm

$$\|\Phi\|_\theta = \sup_X \theta^{-\mathrm{diam}X} \sup_{\xi \in \Omega_X} |\Phi(\xi)| < +\infty. \tag{5.25}$$

We shall say that such interactions are *exponentially decreasing*. If Φ is allowed to be complex-valued we obtain, instead of $\mathcal{B}^\theta$, a complex Banach space $\mathcal{B}^\theta_{\mathbf{C}}$ with norm (5.25).

Notice that $\mathcal{B}^\theta \subset \mathcal{B}_1$ and that

$$\|\cdot\|_1 \leqslant (1-\theta)^{-2}\|\cdot\|_\theta.$$

If $\theta \leqslant \theta'$, then $\mathcal{B}^\theta \subset \mathcal{B}^{\theta'}$ and $\|\cdot\|_\theta \geqslant \|\cdot\|_{\theta'}$.

If the set $M(0)$ in the definition of a $\mathbf{Z}$-morphism $F : (\mathbf{Z}, \Omega'_0, (\bar{\Omega}'_{\Lambda'})_{\Lambda' \in \mathcal{F}}) \mapsto (\mathbf{Z}, \Omega_0, (\bar{\Omega}_\Lambda)_{\Lambda \in \mathcal{F}})$ is chosen to be an interval (which is always possible) and $\Phi \in \mathcal{B}^\theta$, then $F^*\Phi \in \mathcal{B}^\theta(\mathbf{Z}, \Omega'_0, (\bar{\Omega})_{\Lambda' \in \mathcal{F}})$. Restriction of $\mathbf{Z}$ to a subgroup $N\mathbf{Z}$ replaces $\Phi \in \mathcal{B}^\theta$ by $\Phi^* \in \mathcal{B}^{\theta'}$ where $\theta' = \theta^N$, as one readily checks. Therefore one can apply Theorems 5.2 and 5.3 without leaving the realm of exponentially decreasing interactions. In view of this we shall often restrict our attention to mixing $\mathbf{Z}$-lattice systems.

We note for later use that the map $\Phi \mapsto F^{(\beta)*}\Phi^*$ of Theorem 5.3 is linear continuous $\mathcal{B}^\theta(\Omega_0, t) \mapsto \mathcal{B}^{\theta^N}(\Omega_0^{(\beta)}, t^{(\beta)})$ and extends to a $\mathbf{C}$-linear continuous map $\mathcal{B}^\theta_{\mathbf{C}}(\Omega_0, t) \mapsto \mathcal{B}^{\theta^N}_{\mathbf{C}}(\Omega_0^{(\beta)}, t^{(\beta)})$.

5.19 The space $\mathcal{F}^\theta$ and related spaces

If $A \in \mathcal{C}$ [respectively $A \in \mathcal{C}(\Omega_>)$], define

$$\mathrm{var}_n A = \sup\{|A(\xi) - A(\xi')| : \xi_x = \xi'_x \text{ for } |x| \leqslant n\}$$

[respectively

$$\mathrm{var}_n A = \sup\{|A(\xi_>) - A(\xi'_>)| : \xi_x = \xi'_x \text{ for } 1 \leqslant x \leqslant n\}].$$

We denote by $\mathcal{F}^\theta$ [respectively $\mathcal{F}^\theta_>$] the subspace of $\mathcal{C}$ [respectively $\mathcal{C}(\Omega_>)$] consisting of those A such that

$$\|A\|_\theta = \sup_{n \geqslant -1} (\theta^{-2n-1}\mathrm{var}_n A) < +\infty$$

[respectively

$$\|A\|_\theta = \sup_{n \geqslant 0} (\theta^{-n}\mathrm{var}_n A) < +\infty].$$

These quantities define norms on the quotients F^θ, $F^\theta_>$ of $\mathcal{F}^\theta$ and $\mathcal{F}^\theta_>$ by the subspaces of constant functions. F^θ and $F^\theta_>$ are Banach spaces with respect to these norms. $\mathcal{F}^\theta$ and $\mathcal{F}^\theta_>$ are Banach spaces for the norms

$$|||A|||_\theta = \max(\|A\|, \|A\|_\theta).$$

Using complex rather than real functions, one defines similarly complex Banach spaces $\mathcal{F}^\theta_{\mathbf{C}}$, $\mathcal{F}^\theta_{\mathbf{C}>}$, $F^\theta_{\mathbf{C}}$, $F^\theta_{\mathbf{C}>}$.

5.20 Proposition

The image of $\mathcal{B}^\theta$ by the map $\Phi \mapsto A_\Phi$ is $\mathcal{F}^\theta$.

If $\xi = \xi'_x$ for $|x| \leqslant n$, then

$$A_\Phi(\xi) - A_\Phi(\xi') = \sum_X {}^*[\Phi(\xi'|X) - \Phi(\xi|X)],$$

where the sum extends over those X such that 0 is the "middle" element of X in lexicographic order (Section 3.2), and diam $X > 2n$. Therefore

$$\operatorname{var}_n A_\Phi \leqslant \sum_{k=2n+1}^{\infty} 2\|\Phi\|_\theta \theta^k = \frac{2\|\Phi\|_\theta}{1-\theta} \cdot \theta^{2n+1},$$

so that $A_\Phi \in \mathcal{F}^\theta$.

Conversely, if $A \in \mathcal{F}^\theta$, we can write $A = \sum_{n=0}^\infty A_n$, where $A_n \in \mathcal{C}_{[-n,n]}$, and $\|A_n\| \leqslant \operatorname{var}_{n-1} A$ for $n > 0$. We have then $A = A_\Phi$ if $\Phi \in \mathcal{B}^\theta$ is chosen such that $\Phi(\xi|[-n,n]) = -A_n(\xi)$ and $\Phi(\xi|X) = 0$ if X is not a translate of an interval $[-n,n]$. We note for later use that the map $A \mapsto \Phi$ can be chosen to be linear, and thus to have a $\mathbf{C}$-linear continuous extension $\mathcal{F}^\theta_{\mathbf{C}} \mapsto \mathcal{B}^\theta_{\mathbf{C}}$.

5.21 Theorem

Assume that (Ω_0, t) is transitive. Let A, $A' \in \mathcal{F}^\theta$ and ρ, ρ' be the corresponding equilibrium states. Then $\rho = \rho'$ if and only if there exist $c \in \mathbf{R}$ and $C \in \mathcal{F}^\theta$ such that

$$A' - A = c + C \circ \tau - C. \tag{5.26}$$

We have $A = A_\Phi$, $A' = A_{\Phi'}$ with Φ, $\Phi' \in \mathcal{B}^\theta$ (Proposition 5.20). Therefore we are in the situation of Theorem 5.7 and we only have to prove that if Φ, $\Phi' \in \mathcal{B}^\theta$ and $\rho = \rho'$, then (5.26) holds with $C \in \mathcal{F}^\theta$. For this we use the construction of C in the proof of Theorem 5.7.

If $(\tau^j \eta)_x = (\tau^j \xi)_x$ for $|x| \leqslant r, r \geqslant 0$, then (5.12) can now be replaced by

$$
\begin{aligned}
|B(\tau^j \eta) - B(\tau^j \xi)| &= |A_\Psi(\tau^j \eta) - A_\Psi(\tau^j \xi)| \\
&\leqslant 2 \sum_{s=2r+1}^{\infty} \sup_{\xi \in \Omega_{[0,s]}} |\Psi(\xi)| \leqslant 2\|\Psi\|_\theta \sum_{s=2r+1}^{\infty} \theta^s \\
&= \frac{2\|\Psi\|_\theta}{1-\theta} \theta^{2r+1};
\end{aligned}
$$

and therefore

$$
\begin{aligned}
|C(\eta'') - C(\eta')| &\leqslant \frac{4\|\Psi\|_\theta}{1-\theta} \sum_{r=m}^{\infty} \theta^{2r+1} \\
&= \frac{4\|\Psi\|_\theta}{(1-\theta)(1-\theta^2)} \theta^{2m+1},
\end{aligned}
$$

so that $C \in \mathcal{F}^\theta$.

5.22 Remarks

(a) If (Ω_0, t) is mixing, a condition equivalent to (5.26) is

$$\sum_{x \in \mathbf{Z}} [\rho((A' - A) \cdot \tau^x (A' - A)) - (\rho(A' - A))^2] = 0.$$

See Exercise 5(c).

(b) The map $\delta : C \mapsto C \circ \tau - C$ is continuous from F^θ to a closed subspace of F^θ. The proof of Theorem 5.21 shows that it has a continuous inverse $\delta^{-1} : \delta F^\theta \mapsto F^\theta$.

5.23 Lemma

Let $\Phi \in \mathcal{B}^\theta(\Omega_0, t)$, where (Ω_0, t) is mixing.

(a) *$\psi_>$ and $(\psi_>)^{-1}$ belong to $\mathcal{F}^\theta_>$.*
(b) *If $A \in \mathcal{F}^\theta_>$, then*

$$\mathrm{var}_k \mathcal{S}^n A \leqslant c_1 \mathrm{var}_{n+k} A + c_2 \theta^k \|A\|,$$

where c_1, c_2 are constants (depending on Φ, θ).

First notice that, if A and A' belong to $\mathcal{F}^\theta_>$, so does their product, and

$$\|A \cdot A'\|_\theta \leqslant \|A\| \cdot \|A'\|_\theta + \|A\|_\theta \cdot \|A'\|. \tag{5.27}$$

If $A \in \mathcal{F}^\theta_>$ and $\inf |A(\xi_>)| > 0$, then $A^{-i} \in \mathcal{F}^\theta_>$, and

$$\|A^{-1}\|_\theta \leqslant \|A^{-1}\|^2 \cdot \|A\|_\theta. \tag{5.28}$$

We have

$$\text{var}_k W_{\mathbf{Z}_\leqslant,\mathbf{Z}_>}(\xi_\leqslant \vee \cdot) \leqslant 2\|\Phi\|_\theta \sum_{m=0}^{\infty} \sum_{n=k+1}^{\infty} \theta^{m+n}$$
$$= \frac{2\|\Phi\|_\theta}{(1-\theta)^2}\theta^{k+1}. \tag{5.29}$$

Therefore the definition of $\psi_>$ in Section 5.12 gives $\psi_> \in \mathcal{F}_>^\theta$. From (5.20) and (5.28) we obtain also $(\psi_>)^{-1} \in \mathcal{F}_>^\theta$, proving (a).

From Sections 5.11, 5.12 we have

$$\lambda^{-n}\mathfrak{L}^n(\psi_> \cdot A) = \frac{1}{K}\int \sigma_\leqslant(d\xi_\leqslant)\exp[-W_{\mathbf{Z}_\leqslant,\mathbf{Z}_>}(\xi_\leqslant \vee \cdot)]\cdot$$
$$\times [A \circ \alpha_{\mathbf{Z}} \circ \tau^{-n}(\xi_\leqslant \vee \cdot)]$$

hence, using (5.29),

$$\text{var}_k \lambda^{-n}\mathfrak{L}^n(\psi_> \cdot A) \leqslant \|\psi_>\|\text{var}_{n+k}A + c\theta^k\|A\|,$$

where c is a constant (depending on Φ and θ). From this and (5.27), we obtain (b).

5.24 Proposition

Let $\Phi \in \mathcal{B}^\theta(\Omega_0, t)$, where (Ω_0, t) is mixing. The operator $\mathcal{S}$ maps $\mathcal{F}_>^\theta$ into itself and, passing to the quotient by constant functions, defines a bounded operator S on the Banach space $F_>^\theta$ introduced in Section 5.19. The spectral radius of S is strictly less than 1.

From Lemma 5.23 it follows that $\mathcal{S}$ maps $\mathcal{F}_>^\theta$ into itself. Since $\mathcal{S}1 = 1$ (Section 5.14), we may pass to the quotient by constant functions, obtaining a map $S : F_>^\theta \mapsto F_>^\theta$. Part (b) of the lemma gives

$$\text{var}_k S^n A \leqslant c_1 \text{var}_{n+k} A + c_2\theta^k \text{var}_0 A. \tag{5.30}$$

On the other hand, since

$$\{A \in \mathcal{C}(\Omega_>) : |||A|||_\theta \leqslant 1\}$$

is compact in $\mathcal{C}(\Omega_>)$, Proposition 5.16 shows that, given $\delta > 0$, one may choose n such that

$$\text{var}_0 S^n A \leqslant \delta\|A\|_\theta. \tag{5.31}$$

From (5.30) and (5.31) we obtain

$$\begin{aligned}\operatorname{var}_k S^{2n} A &= c_1 \operatorname{var}_{n+k} S^n A + c_2 \theta^k \operatorname{var}_0 S^n A \\ &\leqslant c_1(c_1 \operatorname{var}_{2n+k} A + c_2 \theta^{n+k} \operatorname{var}_0 A) + c_2 \theta^k \delta \|A\|_\theta \\ &\leqslant (c_1^2 \theta^{2n} + c_1 c_2 \theta^n + c_2 \delta) \|A\|_\theta \theta^k;\end{aligned}$$

hence

$$\|S^{2n} A\|_\theta \leqslant \left(c_1^2 \theta^{2n} + c_1 c_2 \theta^n + c_2 \delta\right) \|A\|_\theta .$$

If δ has been chosen sufficiently small and n sufficiently large, the operator norm of $S^{2n} : F_>^\theta \mapsto F_>^\theta$ is strictly less than 1. Therefore the spectral radius of S is strictly less than 1.

5.25 Remark

That the spectral radius of S is strictly less than 1 implies "exponential decrease of correlations"; see Exercise 4(c).

5.26 Theorem

Assume that (Ω_0, t) is transitive. The function $\Phi \mapsto P^\Phi$ on $\mathcal{B}^\theta$ is real analytic.

We first assume that (Ω_0, t) is mixing. If $\Phi \in \mathcal{B}^\theta_{\mathbf{C}}$, the formula*

$$(\mathfrak{L}A)(\xi_>) = \sum_{\xi_0} A(\tau^{-1}(\xi_0 \vee \xi_>)) \exp[-U_{\{0\}}(\xi_0) - W_{\{0\},\mathbf{Z}_>}(\xi_0 \vee \xi_>)]$$

defines a bounded operator $\mathfrak{L}$ on $\mathcal{F}^\theta_{\mathbf{C}>}$, and $\Phi \mapsto \mathfrak{L}$ *is an entire analytic function from $\mathcal{B}^\theta_{\mathbf{C}}$ to the bounded operators on $\mathcal{F}^\theta_{\mathbf{C}>}$*. To see this notice that, for each ξ_0,

$$\Phi \mapsto \{\xi_> \mapsto U_{\{0\}}(\xi_0) + W_{\{0\},\mathbf{Z}_>}(\xi_0 \vee \xi_>)\}$$

is $\mathbf{C}$-linear and continuous, and hence analytic, $\mathcal{B}^\theta_{\mathbf{C}} \mapsto \mathcal{F}^\theta_{\mathbf{C}>}$. Since $\exp : \mathcal{F}^\theta_{\mathbf{C}>} \mapsto \mathcal{F}^\theta_{\mathbf{C}>}$ is analytic, also

$$\Phi \mapsto \{\xi_> \mapsto \exp[-U_{\{0\}}(\xi_0) - W_{\{0\},\mathbf{Z}_>}(\xi_0 \vee \xi_>)]\}$$

is analytic $\mathcal{B}^\theta_{\mathbf{C}} \mapsto \mathcal{F}^\theta_{\mathbf{C}>}$. From this and (5.27) the analyticity of $\Phi \mapsto \mathfrak{L}$ follows.

* Cf. Section 5.11. The various Banach spaces used here are defined in Sections 5.18 and 5.19.

If Φ is real, i.e. $\Phi \in \mathcal{B}^\theta$, we have

$$\mathcal{L}A = \lambda \cdot \psi_> \cdot \mathcal{S}(\psi_>^{-1} \cdot A),$$

and Proposition 5.24 shows that the spectrum of $\mathcal{L}$ consists of $\{\lambda\}$ and a set contained in the disk $\{z : |z| \leqslant \lambda_1\}$ with $\lambda_1 < \lambda$. From this it follows by standard arguments that $\Phi \mapsto \lambda$, and therefore $\Phi \mapsto P^\Phi$, extends to a function analytic in a neighborhood of $\mathcal{B}^\theta$ in $\mathcal{B}^\theta_{\mathbf{C}}$.

The general case of transitive (Ω_0, t) reduces to the mixing case by use of Theorem 5.3. We have indeed

$$P^\Phi = N^{-1} P^{F^{(\beta)*}\Phi^*}$$

by Corollary 5.6 (d), and the map $\Phi \mapsto F^{(\beta)*}\Phi^*$ extends to a $\mathbf{C}$-linear continuous, and therefore analytic, map $\mathcal{B}^\theta_{\mathbf{C}}(\Omega_0, t) \mapsto \mathcal{B}^{\theta^N}_{\mathbf{C}}(\Omega_0^{(\beta)}, t^{(\beta)})$ (see Section 5.18).

5.27 Corollary

Assume that (Ω_0, t) is transitive. The function P on $\mathcal{F}^\theta$ is real analytic.

We can indeed write $A \mapsto P(A)$ as

$$A \mapsto \Phi \mapsto P^\Phi = P(A_\Phi) = P(A),$$

where $A \mapsto \Phi$ extends to an analytic map $\mathcal{F}^\theta_{\mathbf{C}} \mapsto \mathcal{B}^\theta_{\mathbf{C}}$ as noted at the end of the proof of Proposition 5.20.

5.28 Zeta functions

Given a $\mathbf{Z}$-lattice system (Ω_0, t) and $A \in \mathcal{C}$, we write

$$\mathrm{Fix}\,\tau^m = \{\xi \in \Omega : \tau^m \xi = \xi\},$$

$$Z_m(A, \tau) = \sum_{x \in \mathrm{Fix}\,\tau^m} \exp \sum_{k=0}^{m-1} A(\tau^k x),$$

and define a power series

$$\zeta_\tau(ze^A) = \exp \sum_{m=1}^{\infty} Z_m(A, \tau) \frac{z^m}{m}. \tag{5.32}$$

Every τ-periodic point is in one of the sets $F^{(\alpha)}\Omega^{(\alpha)}$ of Theorem 5.2. Therefore $\zeta_\tau(ze^A)$ is a product of factors corresponding to transitive $\mathbf{Z}$-lattice systems.

If (Ω_0, t) is transitive, then Theorem 5.3 shows that $Z_m(A) = 0$ unless m is a multiple of N and

$$Z_{nN}(A, \tau) = N Z_n(A + A \circ \tau + \cdots + A \circ \tau^{N-1}|\Omega^{(\beta)}, \tau^N|\Omega^{(\beta)}),$$
$$\zeta_\tau(ze^A) = \zeta_{\tau^N|\Omega^{(\beta)}}(z^N \exp(A + A \circ \tau + \cdots + A \circ \tau^{N-1})|\Omega^{(\beta)}).$$

The quantity $Z_n(A)$ is a *partition function with periodic boundary conditions*, and the methods of Chapter 3 show that, if (Ω_0, t) is mixing,

$$\lim_{m\to\infty} \frac{1}{m} \log Z_m(A, \tau) = P(A).$$

Therefore the power series (5.32) converges for $|z| < \exp[-P(A)]$ to a holomorphic function of z. In view of Corollary 5.6(c) and (d), this remains true when (Ω_0, t) is not mixing or transitive. It is in fact easily seen that the series (5.32) diverges at $z = \exp[-P(A)]$. Therefore *the radius of convergence of the series* (5.32) is $\exp[-P(A)]$; *this series defines for* $|z| < \exp[-P(A)]$ *a holomorphic function of z, which we call the zeta function (associated with A).*

We shall now show that the domain of analyticity of this function can be extended when A is in a space $\mathcal{F}^\theta$.

5.29 Theorem*

Let (Ω_0, t) be mixing and $A \in \mathcal{F}^\theta$. There exists $R(A) >$ $\exp[-P(A)]$ *such that*

$$d_A(z) = \zeta(ze^A)^{-1} = \exp\left[-\sum_{m=1}^{\infty} \frac{z^m}{m} \sum_{\xi \in \mathrm{Fix}\tau^m} \exp \sum_{k=0}^{m-1} A(\tau^k \xi)\right]$$

extends to an analytic function in $\{z : |z| < R(A)\}$ with only one zero; this zero is simple and located at $\exp[-P(A)]$.

By Proposition 5.20 we have $A = A_\Phi$ with $\Phi \in \mathcal{B}^\theta$. Define $\Phi^{(n)}$ by

$$\Phi^{(n)}(X) = \begin{cases} \Phi(X) & \text{if } \operatorname{diam} X \leqslant n, \\ 0 & \text{if } \operatorname{diam} X > n. \end{cases}$$

Then

$$\|A_\Phi - A_{\Phi^{(n)}}\| \leqslant \sum_{l>n} \|\Phi\|_\theta \theta^l = \|\Phi\|_\theta \frac{\theta^{n+1}}{1-\theta}.$$

* This theorem was announced in Ruelle [6]; see also Exercise 7.

For integers $m \geqslant 1, n \geqslant 0$, we write

$$a(m) = \sum_{\xi \in \mathrm{Fix}\tau^m} \exp \sum_{k=0}^{m-1} A(\tau^k \xi),$$

$$b(m,n) = \sum_{\xi \in \mathrm{Fix}\tau^m} \exp \sum_{k=0}^{m-1} A_{\Phi^{(n)}}(\tau^k \xi)$$

$$= \sum_{\xi \in \mathrm{Fix}\tau^m} \exp \left[- \sum_{k=0}^{m-1} \sum_{l=0}^{n} \Phi(\xi|[k, k+l]) \right].$$

Then

$$\left| \frac{a(m)}{b(m,n)} - 1 \right| \leqslant \exp \left(\frac{\|\Phi\|_\theta}{1-\theta} m\theta^{n+1} \right) - 1.$$

Let $\alpha \in (0, 1)$. If $n = [\alpha m]$ (integer part of αm) we find

$$\left| \frac{a(m)}{b(m,[\alpha m])} - 1 \right| \leqslant \left(\frac{\|\Phi\|_\theta}{1-\theta} m\theta^{\alpha m} \right) \exp \left(\frac{\|\Phi\|_\theta}{1-\theta} m\theta^{\alpha m} \right). \tag{5.33}$$

We have

$$d_A(z) = \exp \left[- \sum_{m=1}^{\infty} b(m, [\alpha m]) \frac{z^m}{m} \right]$$

$$\times \exp \sum_{m=1}^{\infty} [-a(m) + b(m, [\alpha m])] \frac{z^m}{m}.$$

If r is the radius of convergence of the first series in the right-hand side, (5.33) shows that the second series converges for $|z| < r/\theta^\alpha$. Therefore it suffices to prove the theorem for

$$d'(z) = \exp \left[- \sum_{m=1}^{\infty} b(m, [\alpha m]) \frac{z^m}{m} \right]$$

instead of $d_A(z)$.

An operator $L_{(n)}$ is defined on $\mathcal{C}(\Omega_{[1,n]})$ by

$$(L_{(n)}B)(\xi_1, \ldots, \xi_n)$$

$$= \sum_{\xi_0} B(\xi_0, \ldots, \xi_{n-1}) \exp[-U_{\{0\}}(\xi_0) - W_{\{0\},\{1,\ldots,n\}}(\xi_0 \vee \xi_1 \vee \cdots \vee \xi_n)].$$

The trace of an operator on the finite-dimensional space $\mathcal{C}(\Omega_{[1,m]})$ is defined in the usual way, and one checks readily that $\mathrm{tr} L_{(n)}^m = b(m, n)$. This quantity can thus be estimated in terms of the eigenvalues of $L_{(n)}$. Notice that every eigenvalue of $L_{(n)}$ is an eigenvalue of the operator $\mathfrak{L}_{(n)} : \mathcal{F}^{\theta}_{\mathbf{C}>} \mapsto \mathcal{F}^{\theta}_{\mathbf{C}>}$ defined to be the operator $\mathfrak{L}$ for the interaction $\Phi^{(n)}$. But we know from the proof of Theorem 5.26 that $\Phi \mapsto \mathfrak{L}$ is entire analytic (on $\mathcal{B}^{\theta'}_{\mathbf{C}}$, where we choose $\theta' > \theta$). In particular there exists $r' > 0$ such that, for sufficiently large n, the spectrum of $\mathfrak{L}_{(n)}$ has only one eigenvalue larger than r'^{-1} in absolute value. This eigenvalue $\lambda_{(n)}$ is positive and simple, and tends to $\lambda = \exp P^{\Phi}$ when $n \to \infty$. For sufficiently large n we have

$$|b(m,n) - \lambda^m_{(n)}| < |\Omega_{[1,n]}| \cdot r'^{-m},$$

and we can write

$$d'(z) = \exp\left[-\sum_{m=1}^{\infty} \frac{1}{m}(\lambda_{([\alpha m])} z)^m\right] \times \exp \sum_{m=1}^{\infty} [-b(m,[\alpha m]) + \lambda^m_{([\alpha m])}] \frac{z^m}{m}.$$

The radius of convergence of the first series is λ^{-1}, and that of the second series is $\geqslant r'/|\Omega_0|^{\alpha}$. Taking α such that $r'/|\Omega_0|^{\alpha} > \lambda^{-1}$, it suffices to prove the theorem for

$$d''(z) = \exp\left[-\sum_{m=1}^{\infty} \frac{(\lambda z)^m}{m}\right] \times \exp \sum_{m=1}^{\infty} [-\lambda^m_{([\alpha m])} + \lambda^m] \frac{z^m}{m}.$$

The first factor is $\exp\log(1-\lambda z) = 1 - \lambda z$. To study the second factor we estimate

$$\left|\lambda^m_{([\alpha m])} - \lambda^m\right| \leqslant m \cdot \left[\max\left(\lambda, \lambda_{([\alpha m])}\right)\right]^{m-1} \cdot \left|\lambda_{([\alpha m])} - \lambda\right|.$$

By Theorem 5.26, the function $\Phi \mapsto \lambda$ is analytic on $\mathcal{B}^{\theta}$. Replacing θ by θ', where $\theta < \theta' < 1$, we find that

$$|\lambda_n - \lambda| \leqslant C_{\theta'} \|\Phi - \Phi^{(n)}\|_{\theta'},$$
$$\|\Phi - \Phi^{(n)}\|_{\theta'} \leqslant (\theta/\theta')^{n+1} \|\Phi\|_{\theta}.$$

Therefore the radius of convergence of the series

$$\sum_{m=1}^{\infty} [\lambda^m_{([\alpha m])} - \lambda^m] \frac{z^m}{m}$$

is at least $[\lambda \cdot (\theta/\theta')^{\alpha}]^{-1}$, or since we can let θ' tend to 1, $1/\lambda\theta^{\alpha}$. This concludes the proof of the theorem.

5.30 Remark

The function $z \mapsto \zeta(ze^A)$ does not in general extend to a meromorphic function in **C**. Gallovotti [2] has constructed counterexamples of the type discussed in Exercise 8.

Notes

The main fact of the statistical mechanics of one-dimensional systems is the "absence of phase transitions." In particular there is only one Gibbs state: this is true for mixing systems, and interactions in a suitable space $\mathcal{B}_1$ (see Corollary 5.6(b)). The uniqueness of the Gibbs state was proven independently by Ruelle [2] using the transfer matrix and Dobrushin [3] using different methods. Here we generalize Dobrushin's idea, and establish the structure of Gibbs states without the mixing assumption in Theorems 5.2, 5.3. Notice that the case of interactions with finite range reduces to the theory of Markov processes, and had been understood by physicists for a long time. Notice also that mixing systems with "long-range" interactions ($\Phi \in \mathcal{B}\backslash\mathcal{B}_1$) may have several Gibbs states, as seen in Section 3.21*.

The curious Theorem 5.7, and its specialized forms Theorem 5.21, Corollary 7.10(c), Remark 7.11, come from Livšic [1], [2], and Sinai [4]; see also Bowen [6].

The transfer matrix associated with an interaction $\Phi \in \mathcal{B}_1$ is the adjoint of an operator $\mathfrak{L}$ which satisfies a Perron–Frobenius-type theorem† : $\mathfrak{L}$ has a positive eigenvalue equal to the spectral radius. This positive eigenvalue is $\exp P^{\Phi}$; furthermore the spectral properties of $\mathfrak{L}$ are related to cluster properties of the Gibbs state and to analyticity properties of the zeta function. These facts justify the study of $\mathfrak{L}$, which is done here by a new method.

The uniqueness of the Gibbs state is one aspect of the "absence of phase transitions" in one-dimensional systems. Another aspect is constituted by real analyticity properties of the pressure P restricted to suitable subspaces of interactions. Here we consider exponentially decreasing interactions, and prove analyticity of P by showing that $\exp P^{\Phi}$ is an isolated eigenvalue of $\mathfrak{L}$ (we use an idea already present in a paper by Araki [1] on one-dimensional quantum spin systems; see Sinai [4], and Ruelle [5], Appendix B). A zeta function

* In Section 3.21 we followed Israel [1], who proves the existence of interactions with several Gibbs states without explicitly constructing them. Simple examples were known earlier from Dyson [1]. Other examples are obtained in a model due to Fisher [1] and sketched in Exercise 8.

† See for instance Gantmaher [1].

is introduced in Section 5.28: it counts the τ-periodic points of Ω with the weights usual in statistical mechanics. This function has a pole at $\exp[-P^\Phi]$ corresponding to the eigenvalue $\exp P^\Phi$ of $\mathfrak{L}$. A number of other properties of systems with exponentially decreasing interactions are given in the exercises (in particular the exponential decrease of correlations); these properties will be useful in the study of Smale spaces in Chapter 7.

While a strong condition on the interaction, like exponential decrease, seems necessary to obtain that $\exp P^\Phi$ is an isolated eigenvalue of $\mathfrak{L}$, analyticity properties of P can be obtained under less restrictive assumptions. This is a beautiful and somewhat unexpected result of Dobrushin [4]; unfortunately the proof is difficult, and the conditions imposed on the lattice system are probably too stringent. We therefore refer the reader to the original paper for the statement and proof of Dobrushin's theorem.

Exercises

1. Let Ω_0 be a finite set, and $t \neq 0$ a square matrix indexed by $\Omega_0 \times \Omega_0$ such that t_{ij} is an integer $\geqslant 0$ for all i, j. Define $\Omega_0^* = \{(i, j, m) \in \Omega_0 \times \Omega_0 \times \mathbf{Z}_> : m \leqslant t_{ij}\}$ and $t^*_{(i,j,m)(k,l,n)} = 1$ if $j = k$ and $= 0$ otherwise. Then (Ω_0^*, t^*) is a $\mathbf{Z}$-lattice system, which we also denote by $(\Omega_0, t)^*$ (we assume that t is such that $\Omega^* \neq \emptyset$).
   (a) Suppose that t_{ij} takes only the values 0 and 1, so that (Ω_0, t) is a $\mathbf{Z}$-lattice system. If $\xi^* \in \Omega^*$ and $\xi_x^* = (i, j, 1)$, define $(F\xi^*)_x = i$. Show that $F : (\Omega_0^*, t^*) \mapsto (\Omega_0, t)$ is a $\mathbf{Z}$-isomorphism.
   (b) Show that the $\mathbf{Z}$-lattice system obtained by restriction of $\mathbf{Z}$ to $N\mathbf{Z}$, and use of the group isomorphism $\mathbf{Z} \mapsto N\mathbf{Z}$, is $\mathbf{Z}$-isomorphic to $(\Omega_0, t^N)^*$. (Cf. Williams [1].)
2. Let (Ω_0, t) be a mixing $\mathbf{Z}$-lattice system and $\sigma \in \boldsymbol{I}$. If $\eta_0 \vee \eta_> \in \Omega_\geqslant$, let $p_\sigma(\eta_0 \vee \eta_>)$ be the conditional probability that $\xi|\{0\} = \eta_0$ knowing that $\xi|\mathbf{Z}_> = \eta_>$. Let also $g_\sigma(\xi) = -\log p_\sigma(\xi_o \vee \xi_>)$. Then g_σ is defined σ-almost everywhere, $g_\sigma \geqslant 0$, and it is known that

$$\int g_\sigma d\sigma = s(\sigma)$$

[see Billingsley [1], Section 13].

If $\Phi \in \mathcal{B}_1$ and σ is the corresponding Gibbs state, show that g_σ extends to a continuous function on Ω, again denoted by g_σ, such that

$$g_\sigma(\xi) = -\log \frac{\psi_>(\tau^{-1}(\xi_0 \vee \xi_>)) \exp[-U_{\{0\}}(\xi_0) - W_{\{0\},\mathbf{Z}_>}(\xi_0 \vee \xi_>)]}{\lambda \cdot \psi_>(\xi_>)}$$

Show that $\int g_{\sigma'} d\sigma' \leqslant \int g_\sigma d\sigma'$ for all $\sigma' \in I$, with equality only if $\sigma = \sigma'$. [See Parry [2], and also Keane [1], Ledrappier [2].]

3. Let $\Phi, \Phi' \in \mathcal{B}_1(\Omega_0, t)$, where (Ω_0, t) is mixing, and σ, σ' be the corresponding Gibbs states. Prove the following.
   (a) If $A \in \mathcal{C}(\Omega_>)$ and $n \geqslant 0$ then

$$\sigma(A \circ \alpha_{\mathbf{Z}_>}) = \sigma((\mathcal{S}^n A) \circ \alpha_{\mathbf{Z}_>}).$$

   If $A, B \in \mathcal{C}(\Omega_>)$, then

$$\sigma((A \circ \alpha_{\mathbf{Z}_>}) \cdot (B \circ \alpha_{\mathbf{Z}_>} \circ \tau^n)) = \sigma(((\mathcal{S}^n A) \circ \alpha_{\mathbf{Z}_>}) \cdot (B \circ \alpha_{\mathbf{Z}_>})).$$

   (b) Write

$$Z_{n+1}^- = \sigma'\left(\exp \sum_{x=-n}^{0} A_{\Phi-\Phi'} \circ \tau^x\right),$$

$$Z_n^+ = \sigma\left(\exp \sum_{x=1}^{n} A_{\Phi'-\Phi} \circ \tau^x\right),$$

   and define

$$\begin{aligned}\tilde{\sigma} &= \operatorname*{vaguelim}_{n\to+\infty} (Z_{n+1}^-)^{-1}\left(\exp \sum_{x=-n}^{0} A_{\Phi-\Phi'} \circ \tau^x\right) \cdot \sigma' \\ &= \operatorname*{vaguelim}_{n\to+\infty} (Z_n^+)^{-1}\left(\exp \sum_{x=1}^{n} A_{\Phi'-\Phi} \circ \tau^x\right) \cdot \sigma.\end{aligned}$$

   Show that

$$\tau\tilde{\sigma} = \tilde{\sigma} \cdot \exp[A_{\Phi'} - A_\Phi - P^{\Phi'} + P^\Phi].$$

   [Part (b) is due to Sinai: see Gurevič and Oseledec [1]. To prove the existence and equality of the limits use Lemma 5.9(a). Clearly $\tau\tilde{\sigma} = C \exp A_{\Phi'-\Phi} \cdot \tilde{\sigma}$, where $C = \lim_{n\to\infty} Z_{n+1}^-/Z_n^- = \lim(Z_n^-)^{1/n} = \exp(P^\Phi - P^{\Phi'})$ by Proposition 4.4.]

4. Let $\Phi \in \mathcal{B}^\theta(\Omega_0, t)$ where (Ω_0, t) is mixing, and σ be the corresponding Gibbs state. Prove the following.
   (a) If $0 \leqslant A \in \mathcal{C}(\Omega_{[1,m]})$, then $\|\mathcal{S}^m(A \circ \alpha_{[1,m],\mathbf{Z}_>}\|_\theta \leqslant c\sigma(A \circ \alpha_{[1,m]})$ for some c independent of A, m.
   (b) If $0 \leqslant A \in \mathcal{C}(\Omega_{[1,m]})$ and $\sigma(A \circ \alpha_{[1,m]}) > 0$, then

$$\left\| \frac{\mathcal{S}^{m+n}(A \circ \alpha_{[l,m],\mathbf{Z}_>})}{\sigma(A \circ \alpha_{[1,m]})} - 1 \right\| < e^{a-bn},$$

   with real a, b independent of A, m, and $b > 0$.

(c) (Exponential decrease of correlations.) If $0 \leqslant A \in \mathcal{C}_{[-m+1,0]}$ and $0 \leqslant B \in \mathcal{C}_{[n+1,n+m]}$, then

$$|\sigma(AB) - \sigma(A)\sigma(B)| \leqslant e^{a-bn}\sigma(A)\sigma(B).$$

(d) If $A \in \mathcal{C}_{(-\infty,0]}$ and $B \in \mathcal{C}_{[n+1,+\infty)}$, then

$$|\sigma(AB) - \sigma(A)\sigma(B)| \leqslant e^{a-bn}\sigma(|A|)\sigma(|B|).$$

(e) If $b' < \min(b, 2|\log\gamma|)$, there exists a' real such that, if $A, B \in \mathcal{F}^{\gamma}$,

$$|\sigma(A \cdot (B \circ \tau^x)) - \sigma(A)\sigma(B)| \leqslant e^{a'-b'|x|}\|A\|_\gamma \|B\|_\gamma.$$

For fixed θ and $M > 0$ one can choose a', b' such that this formula holds for all Φ with $\|\Phi\|_\theta \leqslant M$.

[(a) follows from the proof of Lemma 5.23, and one can choose c to be a continuous function of $\|\Phi\|_\theta$ for fixed θ. From (a) and Proposition 5.24 one obtains (b), and one can choose a, b, to be continuous functions of $\|\Phi\|_\theta$ (to see this use analyticity with respect to $\Phi \in \mathcal{B}_{\theta'}$ where $\theta < \theta' < 1$, and the fact that $\{\Phi : \|\Phi\|_\theta \leqslant M\}$ is compact in $\mathcal{B}_{\theta'}$). From (b) and exercise 3(a), one obtains (c) and (d). To prove (e) write $A = \Sigma A_n$, $B = \Sigma B_n$ as in the proof of Proposition 5.20.]

5. Let (Ω_0, t) be mixing and $\sigma = \sigma_A$ be the unique equilibrium state for $A \in \mathcal{F}^\theta$. If $B_1, \ldots, B_l \in \mathcal{F}^\theta$, write

$$D^l(B_1, \ldots, B_l) = \frac{\partial^l}{\partial s_1 \cdots \partial s_l} P\left(A + \sum_i s_i B_i\right)\Bigg|_{s_1=\cdots=s_l=0}.$$

(a) $D^1(B_1) = \sigma(B_1)$.
(b) $D^2(B_1, B_2) = \Sigma_{x\in\mathbf{Z}}[\sigma(B_1 \cdot (B_2 \circ \tau^x)) - \sigma(B_1)\sigma(B_2)]$.
(c) $B_1 \mapsto D^2(B_1, B_1)$ is a positive semi-definite quadratic form on $\mathcal{F}^\theta$. Its kernel is $\{c + C\circ\tau - C : c \in \mathbf{R}, C \in \mathcal{F}^\theta\}$ and is thus independent of A. There is $R > 0$ such that $[D^2(B_1, B_1)]^{1/2} \leqslant R\|B_1\|_\theta$.
(d) For all $p \in \mathbf{R}$ (mod 2π),

$$\sum_{x\in\mathbf{Z}} e^{-ipx}[\sigma(B_1 \cdot (B_1 \circ \tau^x)) - \sigma(B_1)\sigma(B_1)] \geqslant 0.$$

[Using Exercise 4(e), it suffices to prove (b) when $A, B_1, B_2 \in \mathcal{C}_\Lambda$ for finite Λ. Going over to a $\mathbf{Z}$-isomorphic system (see the proof of Corollary 4.10(b)), one can assume that

$$A = \tilde{A} \circ \alpha_{\{0\}}, \quad B_1 = \tilde{B}_1 \circ \alpha_{\{0\}}, \quad B_2 = \tilde{B}_2 \circ \alpha_{\{0\}},$$

where $\tilde{A}, \tilde{B}_1, \tilde{B}_2 \in \mathcal{C}(\Omega_0)$. Let $U(\xi_0) = \exp[\tilde{A}(\xi_0) + s\tilde{B}_2(\xi_0)]$ and define the probability measure $\sigma_{n,s}$ on $\Omega_{[-n,n]}$ such that $\sigma_{n,s}\{(\xi_{-n}, \ldots, \xi_n)\}$ is proportional to

$$U(\xi_{-n})t_{\xi_{-n}\xi_{-n+1}}U(\xi_{-n+1})\cdots t_{\xi_{n-1}\xi_n}U(\xi_n).$$

Then the thermodynamic limit of $\sigma_{n,s}$ is σ_{A+sB_2} by uniqueness of the Gibbs state. One finds

$$\frac{d}{ds}\sigma_{n,s}(\tilde{B}_1 \circ \alpha_{\{0\},[-n,n]}) = \sum_{x=-n}^{n} [\sigma_{n,s}((\tilde{B}_1 \circ \alpha_{\{0\},[-n,n]})(\tilde{B}_2 \circ \alpha_{\{x\},[-n,n]})) - \sigma_{n,s}(\tilde{B}_1 \circ \alpha_{\{0\},[-n,n]})\sigma_{n,s}(\tilde{B}_2 \circ \alpha_{\{x\},[-n,n]})].$$

The expression in square brackets decreases exponentially with $|x|$, uniformly in n (and in s for small s). Therefore

$$\lim_{n\to\infty} \frac{d}{ds}\sigma_{n,s}(\tilde{B}_1 \circ \alpha_{\{0\},[-n,n]}) = \sum_{x\in\mathbf{Z}} [\sigma_{A+sB_2}(B_1 \cdot (B_2 \circ \tau^x)) - \sigma_{A+sB_2}(B_1)\sigma_{A+sB_2}(B_2)];$$

hence

$$\frac{d}{ds}\sigma_{A+sB_2}(B_1) = \sum_{x\in\mathbf{Z}} [\sigma_{A+sB_2}(B_1 \cdot (B_2 \circ \tau^x)) - \sigma_{A+sB_2}(B_1)\sigma_{A+sB_2}(B_2)], \tag{*}$$

and (b) follows. That $D^2(B_1, B_1) \geqslant 0$ follows from the convexity of P. In particular $D^2(B_1, B_1) = 0$ implies $D^2(B_1, B_2) = 0$. By induction on l, (*) permits the calculation of D^l, and one checks that $D^2(B_1, B_1) = 0$ implies $D^l(B_1, \ldots, B_l) = 0$ for $l \geqslant 2$. Therefore $d^l\sigma_{A+sB_1}(B_2)/ds^l$ vanishes for $s = 0, l \geqslant 1$, and the function $s \mapsto \sigma_{A+sB_1}(B_2)$, being real analytic, is constant. The determination of the kernel of $B_1 \mapsto D^2(B_1, B_1)$ follows then from Theorem 5.21. The estimate $D^2(B_1, B_2) \leqslant R^2(\|B_1\|_\theta)^2$ follows for instance from Exercise 4(e). To prove (d) notice that there is V unitary in $L^2(\Omega, \sigma)$ such that $V1 = 1$ and $VB = B \circ \tau$. The expression considered is then

$$\sum_{x\in\mathbf{Z}} e^{-ipx}((B_1 - (1, B_1)), V^x(B_1 - (1, B_1)))].$$

6. Let (Ω_0, t) be a $\mathbf{Z}$-lattice system.
   (a) If A is complex continuous on Ω, the radius of convergence of

$$\zeta_\tau(ze^A) = \exp \sum_{m=1}^{\infty} \frac{z^m}{m} \sum_{\xi\in\mathrm{Fix}\tau^m} \exp \sum_{k=0}^{m-1} A(\tau^k\xi)$$

   (or its inverse) is at least $\exp[-P(\mathrm{Re}A)]$.

(b) The functions $A \mapsto \zeta_\tau(e^A)$, $A \mapsto \zeta_\tau(e^A)^{-1}$ are holomorphic in the "convex tube" $\{A : P(\text{Re}A) < 0\}$ of the Banach space of complex continuous functions on Ω.

7. Let (Ω_0, t) be mixing. There is a real continuous function R on $\mathcal{F}^\theta_{\mathbf{C}}$ such that
   (a) $R(A) \geqslant \exp[-P(\text{Re}A)]$.
   (b) If $d_A(z) = \exp[-\sum_{m=1}^\infty \frac{z^m}{m} \sum_{\xi \in \text{Fix}\tau^m} \exp \sum_{k=0}^{m-1} A(\tau^k \xi)]$, then the function $(z, A) \mapsto d_A(z)$ is analytic on

$$\{(z, A) \in \mathbf{C} \times \mathcal{F}^\theta_{\mathbf{C}} : |z| < R(A)\}.$$

   (c) If A is real, then $R(A) > e^{-P(A)}$ and $d_A(\cdot)$ has only one zero in $\{z : |z| < R(A), \}$. This zero is simple and located at $\exp[-P(A)]$.

   [(c) follows from Theorem 5.29, with R continuous on $\mathcal{F}^\theta$ and $R > e^{-P}$: let

$$\beta = \frac{\log r' + P(A)}{\log |\Omega_0| - \log \theta};$$

according to the proof one can take

$$R(A) = \begin{cases} e^{-P(A)}/\theta^\beta = r'/|\Omega_0|^\beta & \text{if} \quad \beta \leqslant 1, \\ e^{-P(A)}/\theta & \text{if} \quad \beta \geqslant 1. \end{cases}$$

From Exercise 6, the function $(z, A) \mapsto d_A(z)$ is holomorphic on $\{(z, A) \in \mathbf{C} \times \mathcal{F}^\theta_{\mathbf{C}} : |z| < \exp[-P(\text{Re}A)]\}$. If $A \in \mathcal{F}^\theta_{\mathbf{C}}$, write $A = B + iC$, where B, $C \in \mathcal{F}^\theta$. We have bounds on the derivatives

$$\frac{d^n}{dz^n} d_{B+tC}(0) \qquad (*)$$

when t is real (corresponding to Theorem 5.29), and when t is not real (corresponding to Exercise 6) as seen above. Using conformal mapping and the formula

$$\frac{1}{2\pi} \int_0^{2\pi} \log |f(\text{re}^{i\theta})| d\theta \geqslant \log |f(0)|$$

(Jensen's theorem), we obtain new bounds on (*). Using these one can extend R continuously from $\mathcal{F}^\theta$ to $\mathcal{F}^\theta_{\mathbf{C}}$ so that (a) and (b) hold.]

8. Let $\Omega_0 = \{0, 1\}$, define an interaction Φ by

$$\Phi(\xi|X|) = \begin{cases} \varphi_{|X|} & \text{if } X \text{ is an interval and } \xi_x = 1 \text{ for all } x \in X, \\ 0 & \text{otherwise;} \end{cases}$$

and write

$$W_n = \sum_{k=1}^n (n - k + 1)\varphi_k, \qquad C = \sum_{k=1}^\infty \varphi_k.$$

Verify that

$$\sum_{m=1}^{\infty} z^m \sum_{\xi \in \mathrm{Fix}\tau^m} \exp \sum_{k=0}^{m-1} A_\Phi(\tau^k \xi) = -z \frac{d}{dz} \left[\log(1-z) + \log(1 - ze^{-C}) + \log\left(1 - \frac{z}{1-z} \sum_{k=1}^{\infty} z^k e^{-W_k}\right) \right];$$

and therefore

$$\zeta(ze^A)^{-1} = (1 - ze^{-C}) \left[1 - z \left(1 + \sum_{k=1}^{\infty} z^k e^{-W_k} \right) \right].$$

[The potential introduced here corresponds to the model of Fisher [1]; see Gallavotti [2].]

9. Let $\Phi \in \mathcal{B}^\theta(\Omega_0, t)$, where (Ω_0, t) is mixing, and σ be the corresponding Gibbs state. Given $A \in \mathcal{F}^\theta$ and a finite interval $\Lambda \subset \mathbf{Z}$, let

$$A_\Lambda(\xi) = |\Lambda|^{-1/2} \sum_{x \in \Lambda} [A(\tau^x \xi) - \sigma(A)],$$

$$D = \sum_{x \in \mathbf{Z}} [\sigma(A \cdot \tau^x A) - \sigma(A)^2] = D^2(A, A) \geqslant 0.$$

Let γ be the gaussian probability measure $(1/\sqrt{2\pi D})e^{-t^2/2D} dt$ on $\mathbf{R}$ if $D > 0$, the unit mass δ_0 at 0 if $D = 0$.

(a) For any integer $n \geqslant 0$

$$\lim_{|\Lambda| \to \infty} \sigma\left(A_\Lambda^{2n}\right) = \frac{(2n)!}{n!} \frac{D^n}{2^n}, \qquad \lim_{|\Lambda| \to \infty} \sigma\left(A_\Lambda^{2n+1}\right) = 0.$$

(b) (Central limit theorem.) When $|\Lambda| \to \infty$, the measure $A_\Lambda \sigma$ tends vaguely to γ. In fact, if $\varphi : \mathbf{R} \to \mathbf{R}$ is continuous with at most polynomial growth, $(A_\Lambda \sigma)(\varphi) \to \gamma(\varphi)$.

(c) More generally, let $A = (A_1, \ldots, A_m) : \Omega \to \mathbf{R}^m$ with $A_1, \ldots, A_m \in \mathcal{F}^\theta$, and write

$$A_\Lambda(\xi) = |\Lambda|^{-1/2} \sum_{x \in \Lambda} [A(\tau^x \xi) - \sigma(A)],$$

$$D_{ij} = \sum_{x \in \mathbf{Z}} [\sigma(A_i \cdot \tau^x A_j) - \sigma(A_i)\sigma(A_j)] = D^2(A_i, A_j).$$

The matrix (D_{ij}) is $\geqslant 0$. Suppose (for simplicity) that it has an inverse (Δ_{ij}), and let γ be the gaussian probability measure

$$(2\pi)^{-m/2} (\det(D_{ij}))^{-1/2} \exp\left(-\frac{1}{2} \sum_{i,j} \Delta_{ij} t_{ij} \right) dt_1 \cdots dt_m.$$

Then, when $|\Lambda| \to \infty$, we have $(A_\Lambda \sigma)(\varphi) \to \gamma(\varphi)$ for all continuous $\varphi : \mathbf{R}^m \mapsto \mathbf{R}$ with at most polynomial growth.

[Notice first that $D = D^2(A, A) \geqslant 0$ by Exercise 5, and $D = 0$ only if $A = c + C \circ \tau - C$ with $c \in \mathbf{R}$, $C \in \mathfrak{F}^\theta$. (a) follows from Exercise 4. Notice that

$$\gamma(t^{2n}) = \frac{(2n)!}{n!} \frac{D^n}{2^n}, \qquad \gamma(t^{2n+1} = 0;$$

hence $(A_\Lambda \sigma)$, $(\varphi) \to \gamma(\varphi)$ when φ is a polynomial; (b) results then from the fact that γ is uniquely determined by its moments. (c) can be easily obtained using Wick's theorem (see for instance Simon [1], Proposition 1.2).]

6

Extension of the thermodynamic formalism

In this chapter we present extensions of some of the results of previous chapters to a more general situation. Proofs will only be sketched, but references to the literature will be given (in the text or in the bibliographic note at the end of the chapter). The extensions consist in replacing the configuration space Ω by a more general compact metrizable space $\mathbf{\Omega}$ on which $\mathbf{Z}^\nu$ acts by homeomorphisms.

6.1 Generalities

Let $\mathbf{\Omega}$ be a non-empty compact metrizable space, and $x \to \tau^x$ a representation of $\mathbf{Z}^\nu$ by homeomorphisms of $\mathbf{\Omega}$ (τ^0 is the identity and $\tau^{x+y} = \tau^x\tau^y$). We denote by $\mathcal{C}$ the Banach algebra $\mathcal{C}(\mathbf{\Omega})$ of real continuous functions on $\mathbf{\Omega}$ with the uniform norm. The probability measures on $\mathbf{\Omega}$ (also called *states*) form a convex compact metrizable subset of the weak dual $\mathcal{C}^*$ of $\mathcal{C}$ ($\mathcal{C}^*$ is the space of real measures on $\mathbf{\Omega}$; the topology of weak dual is the *vague topology*). The set $\boldsymbol{I}$ of τ-invariant states is convex compact; it is a Choquet simplex (see Appendix A.5.5). The extremal points of $\boldsymbol{I}$ are the *ergodic states* and, since $\boldsymbol{I}$ is a metrizable simplex, every $\sigma \in \boldsymbol{I}$ has a unique decomposition into ergodic states: the *ergodic decomposition* (see Appendix A.5.6).

We shall make use of a metric d compatible with the topology of $\mathbf{\Omega}$, but our results will not depend on the special choice of d.

6.2 Expansiveness

The $\mathbf{Z}^\nu$-action τ on $\mathbf{\Omega}$ is called *expansive* if there exists $\varepsilon > 0$ such that

$$(d(\tau^x\xi, \tau^x\eta) \leqslant \varepsilon \quad \text{for all } x \in \mathbf{Z}^\nu) \Rightarrow (\xi = \eta).$$

The number ε is then called an *expansive constant* (with respect to the metric d). It is easily seen that the action τ on Ω considered in Chapter 3 is expansive.

Suppose that τ is expansive with expansive constant ε. Given $\delta > 0$ there exists $L > 0$ such that

$$\begin{aligned} &(d(\tau^x\xi, \tau^x\eta) \leqslant \varepsilon \quad \text{whenever } |x| \leqslant L) \\ &\Rightarrow \quad (d(\xi, \eta) < \delta). \end{aligned} \tag{6.1}$$

[Proof: by compactness.]

6.3 Covers

A family $\mathfrak{A} = (\mathfrak{A}_i)$ of subsets of Ω is a *cover* of Ω if $\cup_i \mathfrak{A}_i = \Omega$; it is a *finite* cover if the index set is finite, an *open* (respectively *Borel, measurable*) cover if the $\mathfrak{A}_i$ are open (respectively Borel, measurable). A *partition* is a cover $(\mathfrak{A}_i)$ such that $i \neq j$ implies $\mathfrak{A}_i \cap \mathfrak{A}_j = \emptyset$.

Let $\mathfrak{A} = (\mathfrak{A}_i)$, $\mathfrak{B} = (\mathfrak{B}_j)$ be covers of Ω. The cover $\mathfrak{A} \vee \mathfrak{B}$ consists of the sets $\mathfrak{A}_i \cap \mathfrak{B}_j$. This notation extends to any finite family of covers. The cover $\tau^{-x}\mathfrak{A}$ consists of the sets $\tau^{-x}\mathfrak{A}_i$, and we write

$$\mathfrak{A}^\Lambda = \bigvee_{x \in \Lambda} \tau^{-x}\mathfrak{A}.$$

We also write

$$\begin{aligned} |\mathfrak{A}| &= \text{number of indices } i \text{ such that } \mathfrak{A}_i \neq \emptyset; \\ \operatorname{diam} \mathfrak{A} &= \sup_i \operatorname{diam} \mathfrak{A}_i, \end{aligned}$$

where $\operatorname{diam} \mathfrak{A}_i$ is the diameter of $\mathfrak{A}_i$ with respect to the metric d.

If $\mathfrak{A} = (\mathfrak{A}_i)$ is an open cover of the compact metrizable set Ω, there is a number $\delta > 0$ (*Lebesgue number*) such that, if $\operatorname{diam} X < \delta$, then $X \subset \mathfrak{A}_i$ for some i. [Proof: by compactness.]

Suppose that τ is expansive with expansive constant ε, and that $\operatorname{diam} \mathfrak{A} \leqslant \varepsilon$. Then, given $\delta > 0$, there exists a finite $\Lambda \subset \mathbf{Z}^\nu$ such that

$$\operatorname{diam} \mathfrak{A}^\Lambda < \delta. \tag{6.2}$$

[This follows from (6.1).]

If $\mathfrak{A}$ is a cover of Ω and $\mathfrak{B}$ a subfamily of $\mathfrak{A}$ which is again a cover of Ω, we say that $\mathfrak{B}$ is a *subcover*; obviously $|\mathfrak{B}| \leqslant |\mathfrak{A}|$.

6.4 Entropy

If $\sigma \in I$ and $\mathfrak{A}$ is a finite Borel partition of Ω, let

$$H(\sigma, \mathfrak{A}) = -\sum_i \sigma(\mathfrak{A}_i) \log \sigma(\mathfrak{A}_i)$$

(where $0\log 0 = 0$ as usual). Then the following exist:

$$\begin{aligned} h_\tau(\sigma, \mathfrak{A}) &= \lim_{\Lambda \nearrow \infty} \frac{1}{|\Lambda|} H(\sigma, \mathfrak{A}^\Lambda) \\ &= \inf_\Lambda \frac{1}{|\Lambda|} H(\sigma, \mathfrak{A}^\Lambda). \end{aligned} \tag{6.3}$$

[Proof: Define $\Omega = (\Omega_0)^{\mathbf{Z}^\nu}$ as in Chapter 3, with $\Omega_0 = \{\mathfrak{A}_i\}$; there is then $\sigma \in I$ such that $H(\sigma, \mathfrak{A}^\Lambda) = S(\alpha_\Lambda \sigma)$ for all finite $\Lambda \subset \mathbf{Z}^\nu$; one then applies Theorem 3.10.] One checks readily that for any nonempty finite $\Lambda \subset \mathbf{Z}^\nu$,

$$h_\tau(\sigma, \mathfrak{A}) = h_\tau(\sigma, \mathfrak{A}^\Lambda). \tag{6.4}$$

The (*mean*) *entropy* of σ is

$$h(\sigma) = h_\tau(\sigma) = \sup_{\mathfrak{A}} h_\tau(\sigma, \mathfrak{A}).$$

This is a real number $\geqslant 0$, or $+\infty$. The entropy is an invariant of the abstract dynamical system (Ω, σ, τ) (the *Kolmogorov-Sinai invariant* for $\nu = 1$).

6.5 Proposition

Let $\sigma \in I$, and $\mathfrak{A}$ be a Borel partition of Ω.

(a) *$h_\tau(\sigma, \mathfrak{A})$ tends to $h(\sigma)$ when* diam $\mathfrak{A} \to 0$. *Therefore h_τ is affine on I.*
(b) *If τ is expansive with expansive constant ε, then $h_\tau(\sigma, \mathfrak{A}) = h(\sigma)$ when* diam $\mathfrak{A} \leqslant \varepsilon$. *In particular, if τ is expansive, h is upper semi-continuous on I.*

[The proof given by Bowen [6], Section 2A, for $\nu = 1$, extends readily. One shows (see Exercise 1) first that

$$h_\tau(\sigma, \mathfrak{A}) - h_\tau(\sigma, \mathfrak{B}) \leqslant H(\sigma, \mathfrak{A} \vee \mathfrak{B}) - H(\sigma, \mathfrak{B}), \tag{6.5}$$

and then that, given $\mathfrak{A}$, the (positive) right-hand side is arbitrarily small for small diam $\mathfrak{B}$. This proves (a). When τ is expansive, (6.2) and (6.4) show that $h_\tau(\sigma, \mathfrak{A}) = h_\tau(\sigma)$ when diam $\mathfrak{A} \leqslant \varepsilon$. Given $\rho \in I$, one can choose $\mathfrak{A}$ such that diam $\mathfrak{A} < \varepsilon$, and the boundaries of the elements $\mathfrak{A}_i$ of the partition $\mathfrak{A}$ have

ρ-measure 0; then the function $\sigma \to H(\sigma, \mathfrak{A}^\Lambda)$ is continuous at ρ for each Λ, and therefore $h_\tau(\cdot) = h_\tau(\cdot, \mathfrak{A})$ is upper semi-continuous at ρ by (6.3).]

6.6 Pressure

Let $\mathfrak{A}$ be any finite cover of Ω. Given $A \in \mathcal{C}$ and a finite $\Lambda \subset \mathbf{Z}^\nu$, we define a *partition function*

$$Z_\Lambda(A, \mathfrak{A}) = \min\left\{ \sum_j \exp\left[\sup_{\xi \in \mathfrak{B}_j} \sum_{x \in \Lambda} A(\tau^x \xi)\right] \right.$$

$$\left. : (\mathfrak{B}_j) \text{ is a subcover of } \mathfrak{A}^\Lambda \right\}.$$

We have

$$Z_{\Lambda+x}(A, \mathfrak{A}) = Z_\Lambda(A, \mathfrak{A}), \tag{6.6}$$

and if $\Lambda_1 \cap \Lambda_2 = \emptyset$, it is readily checked that

$$Z_{\Lambda_1 \cup \Lambda_2}(A, \mathfrak{A}) \leqslant Z_{\Lambda_1}(A, \mathfrak{A}) \cdot Z_{\Lambda_2}(A, \mathfrak{A}). \tag{6.7}$$

We define

$$\begin{aligned} P(A, \mathfrak{A}) &= \lim_{a \to \infty} |\Lambda(a)|^{-1} \log Z_{\Lambda(a)}(A, \mathfrak{A}) \\ &= \inf_a |\Lambda(a)|^{-1} \log Z_{\Lambda(a)}(A, \mathfrak{A}). \end{aligned} \tag{6.8}$$

That the limit exists and is equal to the inf follows from the subadditivity of $a \mapsto \log Z_{\Lambda(a)}(A, \mathfrak{A})$ implied by (6.6), (6.7) (see Appendix A.1.4). We have

$$\exp(-|\Lambda| \cdot \|A\|) \leqslant Z_\Lambda(A, \mathfrak{A}) \leqslant (|\mathfrak{A}| \exp \|A\|)^{|\Lambda|}. \tag{6.9}$$

From the first inequality (6.9) we see that $P(A, \mathfrak{A})$ is finite. Using (6.7) and the second inequality (6.9), we can compare $Z_\Lambda(A, \mathfrak{A})$ with the partition function for a union of disjoint translates of $\Lambda(a)$ contained in Λ. We find thus

$$\limsup_{\Lambda \nearrow \infty} \frac{1}{|\Lambda|} \log Z_\Lambda(A, \mathfrak{A}) = P(A, \mathfrak{A}). \tag{6.10}$$

For finite *open* covers $\mathfrak{A} = (\mathfrak{A}_i)$ of Ω the limit

$$P(A) = P_\tau(A) = \lim_{\operatorname{diam} \mathfrak{A} \to 0} P(A, \mathfrak{A}) \tag{6.11}$$

exists, finite or $+\infty$. [Bowen [6], Proposition 2.8. The idea is that if $\mathfrak{B} = (\mathfrak{B}_j)$ is an open cover such that each $\mathfrak{B}_j$ is contained in some $\mathfrak{A}_i$, then $P(A, \mathfrak{A}) \leqslant P(A, \mathfrak{B}) + \delta$, where δ is the maximum of the oscillation of A on a set $\mathfrak{A}_i$

(i.e. $\delta = \max_i \sup\{|A(\boldsymbol{\xi}) - A(\boldsymbol{\eta})| : \boldsymbol{\xi}, \boldsymbol{\eta} \in \mathfrak{A}_i\}$). This holds in particular if diam $\mathfrak{B}$ is a Lebesgue number for $\mathfrak{A}$.] The limit $P(A)$ defined by (6.11) is called the (*topological*) *pressure* of $A \in \mathcal{C}$.

If we have a sequence (or net) of open covers $\mathfrak{A}$ and, for each $\mathfrak{A}$, a non-empty finite set $\Lambda \subset \mathbf{Z}^\nu$, so that diam $\mathfrak{A}^\Lambda \to 0$, then $P(A, \mathfrak{A}) \to P(A)$. [The proof is similar to that of (6.11): if diam $\mathfrak{B}^\Lambda$ is a Lebesgue number for $\mathfrak{A}$, then

$$Z_{\Lambda(a)}(A, \mathfrak{A}) \leqslant Z_{\Lambda(a)+\Lambda}(A, \mathfrak{B}) \exp[|\Lambda(a)|\delta + (|\Lambda(a) + \Lambda| - |\Lambda(a)|)\|A\|]$$

and therefore $P(A, \mathfrak{A}) \leqslant P(A, \mathfrak{B}) + \delta$.] In particular, if τ is expansive and diam $\mathfrak{A}$ an expansive constant, then $P(A, \mathfrak{A}) = P(A)$.

6.7 Other definitions of the pressure

Let again $A \in \mathcal{C}$ and $\mathfrak{A}$ be a finite open cover of $\boldsymbol{\Omega}$. We define a new partition function

$$Z_\Lambda^{(1)}(A, \mathfrak{A}) = \min\left\{ \sum_j \exp\left[\inf_{\xi \in \mathfrak{B}_j} \sum_{x \in \Lambda} A(\tau^x \boldsymbol{\xi}) \right] \right.$$

$$\left. :(\mathfrak{B}_j) \text{ is a subcover of } \mathfrak{A}^\Lambda \right\}.$$

Then

$$1 \leqslant \frac{Z_\Lambda(A, \mathfrak{A})}{Z_\Lambda^{(1)}(A, \mathfrak{A})} \leqslant e^{|\Lambda|\delta},$$

where δ is the maximum of the oscillation of A on a set of $\mathfrak{A}^\Lambda$, and one checks that

$$P(A) = \lim_{\text{diam}\,\mathfrak{A} \to 0} \limsup_{\Lambda \nearrow \infty} \frac{1}{|\Lambda|} \log Z_\Lambda^{(1)}(A, \mathfrak{A})$$

$$= \sup_{\mathfrak{A}} \limsup_{\Lambda \nearrow \infty} \frac{1}{|\Lambda|} \log Z_\Lambda^{(1)}(A, \mathfrak{A}).$$

One can in this formula replace Λ by $\Lambda(a)$ and $\Lambda \nearrow \infty$ by $a \to \infty$. One also has

$$P(A) = \lim_{\text{diam}\mathfrak{A} \to \infty} \liminf_{a \to \infty} \frac{1}{|\Lambda(a)|} \log Z_{\Lambda(a)}^{(1)}(A, \mathfrak{A}).$$

Let S be a finite subset of $\boldsymbol{\Omega}$. We say that S is (Λ, ε)-*separated* if $\boldsymbol{\xi}, \boldsymbol{\eta} \in S$ and $\boldsymbol{\xi} \neq \boldsymbol{\eta}$ imply

$$d(\tau^x \boldsymbol{\xi}, \tau^x \boldsymbol{\eta}) > \varepsilon \quad \text{for some} \quad x \in \Lambda.$$

We say that S is (Λ, ε)-*spanning* if for all $\eta \in \Omega$ there exists $\xi \in S$ such that

$$d(\tau^x \xi, \tau^x \eta) \leqslant \varepsilon \qquad \text{for all} \quad x \in \Lambda.$$

One can then define partition functions

$$Z_\Lambda^{(2)}(A, \varepsilon) = \sup \left\{ \sum_{\xi \in S} \exp \sum_{x \in \Lambda} A(\tau^x \xi) : S \text{ is } (\Lambda, \varepsilon)\text{-separated} \right\},$$

$$Z_\Lambda^{(3)}(A, \varepsilon) = \inf \left\{ \sum_{\xi \in S} \exp \sum_{x \in \Lambda} A(\tau^x \xi) : S \text{ is } (\Lambda, \varepsilon)\text{-spanning} \right\};$$

and, for $i = 2, 3$,

$$\limsup_{\Lambda \nearrow \infty} \frac{1}{|\Lambda|} \log Z_\Lambda^{(i)}(A, \varepsilon)$$

increases as ε decreases. One shows [Walters [1], Section 1] that

$$\begin{aligned} P(A) &= \lim_{\varepsilon \to 0} \limsup_{\Lambda \nearrow \infty} \frac{1}{|\Lambda|} \log Z_\Lambda^{(i)}(A, \varepsilon) \\ &= \lim_{\varepsilon \to 0} \limsup_{a \to \infty} \frac{1}{|\Lambda(a)|} \log Z_{\Lambda(a)}^{(i)}(A, \varepsilon) \end{aligned}$$

(see Exercise 2). One also has

$$P(A) = \lim_{\varepsilon \to 0} \liminf_{a \to \infty} \frac{1}{|\Lambda(a)|} \log Z_{\Lambda(a)}^{(i)}(A, \varepsilon).$$

If τ is expansive, it is not necessary to let diam $\mathfrak{A} \to 0$ or $\varepsilon \to 0$ in the formulae for the pressure (of this section and the preceding one). It suffices to take for diam $\mathfrak{A}$ or ε an expansive constant. In particular

$$\begin{aligned} P(A) &= \lim_{a \to \infty} \frac{1}{|\Lambda(a)|} \log Z_{\Lambda(a)}(A, \mathfrak{A}) \\ &= \lim_{a \to \infty} \frac{1}{|\Lambda(a)|} \log Z_{\Lambda(a)}^{(1)}(A, \mathfrak{A}) \\ &= \lim_{a \to \infty} \frac{1}{|\Lambda(a)|} \log Z_{\Lambda(a)}^{(i)}(A, \varepsilon), \quad i = 2, 3. \end{aligned}$$

Therefore the definition of the pressure given in Theorem 3.4 is a special case of that given here.

6.8 Properties of the pressure

Either $P(A) = +\infty$ for all $A \in \mathcal{C}$, or $P(A)$ is finite for all $A \in \mathcal{C}$. In the latter case P is convex, and increasing [i.e. $A \leqslant B$ implies $P(A) \leqslant P(B)$] on $\mathcal{C}$; P is also continuous:

$$|P(A) - P(B)| \leqslant ||A - B||,$$

and, if $t \in \mathbf{R}$,

$$P(A + B \circ \tau^x - B + t) = P(A) + t \quad (t \in \mathbf{R}),$$
$$P(A + B) \leqslant P(A) + P(B),$$
$$|P(A)| \leqslant P(|A|).$$

[For the easy proofs, see Walters [1], Theorem 2.1.]

6.9 The action τ^a

Given integers $a_1, \ldots, a_\nu > 0$, write $ax = (a_1 x_1, \ldots, a_\nu x_\nu)$. A $\mathbf{Z}^\nu$-action τ^a is defined by $(\tau^a)^x = \tau^{ax}$, and one checks easily that

$$h_\tau(\sigma, \mathfrak{A}) = \frac{1}{|\Lambda(a)|} h_{\tau^a}\big(\sigma, \mathfrak{A}^{\Lambda(a)}\big), \tag{6.12}$$

$$P_\tau(A, \mathfrak{A}) = \frac{1}{|\Lambda(a)|} P_{\tau^a}\left(\sum_{x \in \Lambda(a)} A \circ \tau^x, \mathfrak{A}^{\Lambda(a)}\right). \tag{6.13}$$

[Bowen [6], Lemma 2.9.]

6.10 Lemma

Let $\mathfrak{B}$ be a finite open cover of Ω, and Λ a finite subset of $\mathbf{Z}^\nu$. There exists a Borel partition $\mathfrak{B}_\Lambda$ of Ω such that

(a) *each set of $\mathfrak{B}_\Lambda$ is contained in some set of $\mathfrak{B}^\Lambda$;*
(b) *each $\xi \in \Omega$ is in the closure of at most $|\Lambda| \cdot |\mathfrak{B}|$ sets of $\mathfrak{B}_\Lambda$.*

[Bowen [6], Lemma 2.12. The idea is as follows. A partition of unity subordinate to $\mathfrak{B}$ gives a map $\alpha : \Omega \mapsto \Delta$ where Δ is a $(|\mathfrak{B}| - 1)$-dimensional simplex. Let $\beta = (\alpha \circ \tau^x)_{x \in \Lambda} : \Omega \mapsto \Delta^\Lambda$; one takes for $\mathfrak{B}_\Lambda$ the inverse image by β of a suitable partition of the $(|\mathfrak{B}| \cdot |\Lambda| - |\Lambda|)$-dimensional set Δ^Λ.]

6.11 Lemma

If $\mathfrak{A}$ is a Borel partition of Ω such that each $\xi \in \Omega$ is in the closure of at most M members of $\mathfrak{A}$, then

$$h(\sigma, \mathfrak{A}) + \sigma(A) \leqslant P(A) + \log M. \tag{6.14}$$

[Bowen [6], Lemma 2.11. Denote by $\mathfrak{A}_i^\Lambda (i = 1, 2, \ldots)$ the non-empty sets in $\mathfrak{A}^\Lambda$, and for each i choose $\xi_i \in \mathfrak{A}_i^\Lambda$ such that

$$\int_{\mathfrak{A}_i^\Lambda} \sigma(d\boldsymbol{\xi}) \sum_{x\in\Lambda} A(\tau^x \xi) \leqslant \sigma(\mathfrak{A}_i^\Lambda) \sum_{x\in\Lambda} A(\tau^x \xi_i).$$

Then

$$\begin{aligned} h_\tau(\sigma, \mathfrak{A}) + \sigma(A) &\leqslant |\Lambda|^{-1} \left[H(\sigma, \mathfrak{A}^\Lambda) + \sigma\left(\sum_{x\in\Lambda} A \circ \tau^x\right)\right] \\ &\leqslant |\Lambda|^{-1} \sum_i \sigma(\mathfrak{A}_i^\Lambda) \left(-\log\sigma(\mathfrak{A}_i^\Lambda) + \sum_{x\in\Lambda} A(\tau^x \xi_i)\right) \\ &\leqslant |\Lambda|^{-1} \log \sum_i \exp \sum_{x\in\Lambda} A(\tau^x \xi_i). \end{aligned}$$

Let $\mathfrak{B}$ be a finite open cover of $\boldsymbol{\Omega}$ such that each member of $\mathfrak{B}$ intersects at most M members of $\mathfrak{A}$, and let $\mathfrak{B}' = (\mathfrak{B}'_j)$ be a subcover of $\mathfrak{B}^\Lambda$. For each $\mathfrak{A}_i^\Lambda$ choose $\mathfrak{B}'_j \ni \boldsymbol{\xi}_i$; this map $\mathfrak{A}_i^\Lambda \to \mathfrak{B}'_j$ is at most $M^{|\Lambda|}$ to one; therefore

$$h_\tau(\sigma, \mathfrak{A}) + \sigma(A) \leqslant |\Lambda|^{-1} \log \sum_j M^{|\Lambda|} \sup_{\xi\in\mathfrak{B}'_j} \sum_{x\in\Lambda} A(\tau^x \boldsymbol{\xi}).$$

Taking the minimum over the subcovers $(\mathfrak{B}'_j)$ we have

$$h_\tau(\sigma, \mathfrak{A}) + \sigma(A) \leqslant \log M + |\Lambda|^{-1} \log Z_\Lambda(A, \mathfrak{B}).$$

Taking $\Lambda \nearrow \infty$, then diam $\mathfrak{B} \to 0$, we obtain (6.14).]

6.12 Theorem (variational principle)

For all $A \in \mathcal{C}$,

$$P(A) = \sup_{\sigma\in \boldsymbol{I}} [h(\sigma) + \sigma(A)].$$

This is the main general theorem on pressure. We give the outline of a proof (assuming for simplicity $P(A)$ finite). In part (a) we follow Bowen ([6], Section 2.B), in part (b) we follow Denker ([1], Theorem 2).

(a) *Given $A \in \mathcal{C}$, we have for all $\boldsymbol{\sigma} \in \boldsymbol{I}$*

$$h(\sigma) + \sigma(A) \leqslant P(A). \tag{6.15}$$

The idea is to eliminate the term $\log M$ in (6.14) by replacing the $\mathbf{Z}^\nu$-action τ by τ^a. Let $\mathfrak{A}$ be a Borel partition of $\mathbf{\Omega}$. By (6.12) we have

$$h_\tau(\sigma, \mathfrak{A}) + \sigma(A) = \frac{1}{|\Lambda(a)|}\left[h_{\tau^a}(\sigma, \mathfrak{A}^{\Lambda(a)}) + \sigma\left(\sum_{x\in\Lambda(a)} A\circ\tau^x\right)\right].$$

Choose now $\mathfrak{B}_{\Lambda(a)}$ according to Lemma 6.10 with any finite open covering $\mathfrak{B}$ with small diam $\mathfrak{B}$. According to (6.5), and Lemma 6.11 applied to $\mathfrak{B}_{\Lambda(a)}$, the above expression is

$$\leqslant \frac{1}{|\Lambda(a)|}\left[h_{\tau^a}(\sigma, \mathfrak{B}_{\Lambda(a)}) + \sigma\left(\sum_{x\in\Lambda(a)} A\circ\tau^x\right)\right.$$
$$\left. + H\left(\sigma, \mathfrak{A}^{\Lambda(a)} \vee \mathfrak{B}_{\Lambda(a)}\right) - H\left(\sigma, \mathfrak{B}_{\Lambda(a)}\right)\right]$$
$$\leqslant \frac{1}{|\Lambda(a)|}\left[P_{\tau^a}\left(\sum_{x\in\Lambda(a)} A\circ\tau^x\right) + \log(|\Lambda(a)|\cdot|\mathfrak{B}|)\right.$$
$$\left. + H\left(\sigma, \mathfrak{A}^{\Lambda(a)} \vee \mathfrak{B}_{\Lambda(a)}\right) - H\left(\sigma, \mathfrak{B}_{\Lambda(a)}\right)\right].$$

This first term is $P(A)$ by (6.13); $|\Lambda(a)|^{-1}\log(|\Lambda(a)|\cdot|\mathfrak{B}|)$ tends to zero when $a\to\infty$; finally one has [see Exercise 1(a)]

$$\frac{1}{|\Lambda(a)|}\left[H\left(\sigma, \mathfrak{A}^{\Lambda(a)} \vee \mathfrak{B}_{\Lambda(a)}\right) - H(\sigma, \mathfrak{B}_{\Lambda(a)})\right]$$
$$\leqslant \frac{1}{|\Lambda(a)|}\sum_{x\in\Lambda(a)}\left[H\left(\sigma, (\tau^{-x}\mathfrak{A}) \vee \mathfrak{B}_{\Lambda(a)}\right) - H(\sigma, \mathfrak{B}_{\Lambda(a)})\right]$$
$$= \frac{1}{|\Lambda(a)|}\sum_{x\in\Lambda(a)}\left[H\left(\sigma, \mathfrak{A} \vee \tau^x\mathfrak{B}_{\Lambda(a)}\right) - H\left(\sigma, \tau^x\mathfrak{B}_{\Lambda(a)}\right)\right]$$

and, since diam $\tau^x\mathfrak{B}_{\Lambda(a)} \leqslant$ diam $\mathfrak{B}$, this is arbitrarily small [see Exercise 1(b)].

(b) *Given $A \in \mathcal{C}$ and $\varepsilon > 0$, there exists $\sigma \in I$ such that*

$$h(\sigma) + \sigma(A) \geqslant P(A) - 3\varepsilon. \tag{6.16}$$

Choose a finite open cover $\mathfrak{B}$ such that

$$P(A, \mathfrak{B}) \geqslant P(A) - \varepsilon$$

and diam $\mathfrak{B}$ is so small that $|A(\boldsymbol{\xi}) - A(\boldsymbol{\eta})| < \varepsilon$ if $\boldsymbol{\xi}, \boldsymbol{\eta} \in \mathfrak{B}_j$. Now choose a such that $|\Lambda(a)|^{-1}\log(|\Lambda(a)|\cdot|\mathfrak{B}|) < \varepsilon$, and choose $\mathfrak{A} = \mathfrak{B}_{\Lambda(a)}$ according to

Lemma 6.10. Writing $\Omega_0 = \{\mathfrak{A}_i\}$, construct the spaces

$$\Omega = \left\{ (\mathfrak{A}_{i_x}) \in (\Omega_0)^{\mathbf{Z}^\nu} : \bigcap_{x \in \mathbf{Z}^\nu} \text{closure } \tau^{ax}\mathfrak{A}_{i_x} \neq \emptyset \right\},$$

$$\Omega^* = \left\{ (\boldsymbol{\xi}, (\mathfrak{A}_{i_x})) \in \Omega \times \Omega : \boldsymbol{\xi} \in \bigcap_{x \in \mathbf{Z}^\nu} \text{closure } \tau^{ax}\mathfrak{A}_{i_x} \right\}.$$

Let $p : \Omega^* \mapsto \boldsymbol{\Omega}$, $q : \Omega^* \mapsto \Omega$ be the projections, and define $B \in \mathcal{C}(\Omega)$ by

$$B(\mathfrak{A}_{i_x}) = \max_{j : \mathfrak{A}_{i_0} \subset \mathfrak{B}_j} \sup_{\xi \in \mathfrak{B}_j} \sum_{x \in \Lambda(a)} A(\tau^x \boldsymbol{\xi}).$$

Then

$$\begin{aligned} P(A) - \varepsilon \leqslant P(A, \mathfrak{B}) &= \frac{1}{|\Lambda(a)|} P_{\tau^a}\left(\sum_{x \in \Lambda(a)} A \circ \tau^x, \mathfrak{B}^{\Lambda(a)} \right) \\ &\leqslant \frac{1}{|\Lambda(a)|} P(B) = \frac{1}{|\Lambda(a)|}[s(\sigma) + \sigma(B)], \end{aligned}$$

where we have used Theorem 3.12 and σ is a translation invariant probability measure on Ω. There exists (by the Hahn–Banach and Markov–Kakutani theorems – see Appendix A.3.2 and A.3.4) a translation invariant probability measure σ^* on Ω^* such that $q\sigma^* = \sigma$. Let $\tilde{\sigma} = p\sigma^*$; then $\tilde{\sigma}$ is τ^a-invariant and

$$\sigma(B) \leqslant \tilde{\sigma}\left(\sum_{x \in \Lambda(a)} A \circ \tau^x \right) + |\Lambda(a)|\varepsilon,$$

$$s(\sigma) \leqslant h_{\tau^a}(\tilde{\sigma}) + \log(|\Lambda(a)| \cdot |\mathfrak{B}|);$$

[estimate $h_{\tau^a}(\tilde{\sigma})$ with the partition formed by the intersections of the closures of the $\mathfrak{A}_i$]. Therefore

$$P(A) - \varepsilon \leqslant \frac{1}{|\Lambda(a)|}\left[h_{\tau^a}(\tilde{\sigma}) + \tilde{\sigma}\left(\sum_{x \in \Lambda(a)} A \circ \tau^x \right) \right] + 2\varepsilon,$$

or writing $\boldsymbol{\sigma} = |\Lambda(a)|^{-1} \Sigma_{x \in \Lambda(a)} \tau^x \tilde{\sigma} \in \boldsymbol{I}$,

$$P(A) - 3\varepsilon \leqslant \frac{1}{|\Lambda(a)|} h_{\tau^a}(\boldsymbol{\sigma}) + \boldsymbol{\sigma}(A) = h_\tau(\boldsymbol{\sigma}) + \boldsymbol{\sigma}(A).$$

6.13 Equilibrium states

Suppose that P is not $+\infty$. The set of *equilibrium states* for $A \in \mathcal{C}$ is by definition

$$\boldsymbol{I}_A = \{\boldsymbol{\sigma} \in \boldsymbol{I} : h(\boldsymbol{\sigma}) + \boldsymbol{\sigma}(A) = P(A)\}.$$

This set can be empty (see for instance Gurevič [1]. In particular, it need not coincide with

$$\boldsymbol{I}'_A = \{\boldsymbol{\sigma} \in \mathcal{C}^* : P(A+B) \geqslant P(A) + \boldsymbol{\sigma}(B) \quad \text{for all} \quad B \in \mathcal{C}\}. \quad (6.17)$$

The interesting case (next theorem) is when h_τ is upper semi-continuous on $\boldsymbol{I}$. This happens if τ is expansive (Proposition 6.5(b)).

6.14 Theorem

Let h_τ be finite upper semi-continuous on $\boldsymbol{I}$.

(a) $\boldsymbol{I}_A = \boldsymbol{I}'_A \neq \emptyset$; *$\boldsymbol{I}_A$ is convex, compact, a Choquet simplex, and a face of $\boldsymbol{I}$.*
(b) *The set*

$$\boldsymbol{D} = \{A \in \mathcal{C} : \boldsymbol{I}_A \text{ consists of a single point}\}$$

is residual in $\mathcal{C}$.
(c) *For all $\boldsymbol{\sigma} \in \boldsymbol{I}$,*

$$h(\boldsymbol{\sigma}) = \inf_{A \in \mathcal{C}} [P(A) - \boldsymbol{\sigma}(A)].$$

As in the proof of Theorem 3.7, one sees that $\boldsymbol{I}'_A$ is convex, compact, and contained in $\boldsymbol{I}$, and that

$$\boldsymbol{D}' = \{A \in \mathcal{C} : \boldsymbol{I}'_A \text{ consists of a single point}\}$$

is residual in $\mathcal{C}$.

Since h_τ is upper semi-continuous, we have $\boldsymbol{I}_A \neq \emptyset$, and the proof of (c) is the same as that of the corresponding statement of Theorem 3.12. The proof of Theorem 3.12 also shows that $\boldsymbol{I}_A = \boldsymbol{I}'_A$. Therefore $\boldsymbol{D} = \boldsymbol{D}'$ is residual in $\mathcal{C}$, proving (b).

Finally the proof that $\boldsymbol{I}_A$ is a Choquet simplex and a face of $\boldsymbol{I}$ is the same as that of Corollary 3.14.

6.15 Remark

Suppose that $P(0) < +\infty$ and that, for all $\boldsymbol{\sigma} \in \boldsymbol{I}$,

$$h(\boldsymbol{\sigma}) = \inf_{A \in \mathcal{C}} [P(A) - \boldsymbol{\sigma}(A)]. \quad (6.18)$$

Then h is upper semi-continuous, as the inf of continuous functions. By Theorem 6.14(c), (6.18) is thus equivalent to the upper semi-continuity of h.

Israel's theorem (Appendix A.3.6) applies to the present situation, permitting the approximation of invariant states by equilibrium states. In particular the following analogs of Theorem 3.16 and Corollary 3.17 hold.

Let h_τ be finite and upper semi-continuous. Given $A \in \mathcal{C}$, $\boldsymbol{\sigma} \in \boldsymbol{I}$, and $\varepsilon > 0$, there exist $A' \in \mathcal{C}$ and $\boldsymbol{\sigma}' \in \boldsymbol{I}_{A'}$ such that

$$\|\boldsymbol{\sigma}' - \boldsymbol{\sigma}\| \leqslant \varepsilon$$

and

$$\|A' - A\| \leqslant \frac{1}{\varepsilon}[P(A) - \boldsymbol{\sigma}(A) - h_\tau(\boldsymbol{\sigma})].$$

The union of the $\boldsymbol{I}_A$ for $A \in \mathcal{C}$, i.e., the set of all equilibrium states, is dense in $\boldsymbol{I}$ for the norm topology.

If $\rho_1, \ldots, \rho_n$ are ergodic states, there is $A \in \mathcal{C}$ such that $\rho_1, \ldots, \rho_n \in \boldsymbol{I}_A$.

6.16 Commuting continuous maps

If, instead of a $\mathbf{Z}^\nu$-action τ on $\boldsymbol{\Omega}$, generated by ν commuting homeomorphisms, we are given ν commuting continuous maps, we can extend most of the previous results of this chapter. This is indicated in Section 6.18. We omit (for simplicity) considerations of expansiveness.

τ^x is now defined only when $x \in \mathbf{Z}^\nu_\geqslant = \{x \in \mathbf{Z}^\nu : x_1, \ldots, x_\nu \geqslant 0\}$; it is thus a $\mathbf{Z}^\nu_\geqslant$-action. We indicate how to associate with it a $\mathbf{Z}^\nu$-action.

6.17 Extension to a $\mathbf{Z}^\nu$-action

First, let

$$\boldsymbol{\Omega}' = \bigcap_{x \in \mathbf{Z}^\nu_\geqslant} \tau^x \boldsymbol{\Omega}. \tag{6.19}$$

The $\mathbf{Z}^\nu_\geqslant$-action τ restricts to $\boldsymbol{\Omega}'$, and $\tau^x : \boldsymbol{\Omega}' \mapsto \boldsymbol{\Omega}'$ is onto for all $x \in \mathbf{Z}^\nu_\geqslant$. We construct now a compact metrizable space $\underline{\boldsymbol{\Omega}}$ with $\mathbf{Z}^\nu$-action τ and a continuous map $\pi : \underline{\boldsymbol{\Omega}} \mapsto \boldsymbol{\Omega}$, such that

$$\pi \underline{\boldsymbol{\Omega}} = \boldsymbol{\Omega}'$$

and $\pi \tau^a = \tau^a \pi$ when $a \in \mathbf{Z}^\nu_\geqslant$.

We let

$$\underline{\boldsymbol{\Omega}} = \{(\xi_x) \in (\boldsymbol{\Omega}')^{\mathbf{Z}^\nu_\leqslant} : \tau^a \xi_{x-a} = \xi_x \text{ if } a \in \mathbf{Z}^\nu_\geqslant \text{ and } x \in \mathbf{Z}^\nu_\leqslant\}.$$

This is compact metrizable as a closed subset of a countable product of compact metrizable sets. We also let

$$\tau^{a-b}(\xi_x) = (\eta_x) \quad \text{with} \quad \eta_x = \tau^a \xi_{x-b}$$

if $a, b \in \mathbf{Z}^{\nu}_{\geqslant}$. The consistency of this definition is easily checked, and yields a $\mathbf{Z}^{\nu}$-action by homeomorphisms τ^x of $\underline{\Omega}$. Finally we write

$$\pi(\xi_x) = \xi_0.$$

It is then clear that $\pi\tau^a = \tau^a\pi$ when $a \in \mathbf{Z}^{\nu}_{\geqslant}$. One also verifies easily that

$$\pi\underline{\Omega} = \Omega'.$$

If σ is a τ-invariant probability measure on Ω, then supp $\sigma \subset \Omega'$. By the Hahn-Banach and Markov-Kakutani theorems* there is a probability measure $\underline{\sigma}$ on $\underline{\Omega}$ such that $\pi\underline{\sigma} = \sigma$, and $\tau^x\underline{\sigma} = \underline{\sigma}$ for all $x \in \mathbf{Z}^{\nu}_{\geqslant}$. Then $\tau^x\underline{\sigma} = \underline{\sigma}$ for all $x \in \mathbf{Z}^{\nu}$ (because the τ^x are homeomorphisms of $\underline{\Omega}$). In fact $\underline{\sigma}$ is unique because the functions $A \circ \pi \circ \tau^x$, with $A \in \mathcal{C}$ and $x \in \mathbf{Z}^{\nu}$, are dense in $\mathcal{C}(\underline{\Omega})$. Therefore $\pi : \underline{\sigma} \mapsto \sigma$ is a bijection of the τ-invariant states on $\underline{\Omega}$ to the τ-invariant states on Ω (it is an affine homeomorphism).

6.18 Results for $\mathbf{Z}^{\nu}_{\geqslant}$-actions

As already mentioned, we omit all considerations of expansiveness. Apart from that the definitions and results of Sections 6.1, 6.3, 6.4, 6.5 carry over to the present case, with the following remarks.

(a) We define and consider $\mathfrak{A}^{\Lambda}$ only when $\Lambda \subset \mathbf{Z}^{\nu}_{\geqslant}$.
(b) If $\sigma \in \boldsymbol{I}$, let $\underline{\sigma}$ be obtained by the construction of Section 6.17. Then

$$H(\sigma, \mathfrak{A}^{\Lambda}) = H(\underline{\sigma}, \pi^{-1}\mathfrak{A}^{\Lambda}) = H(\underline{\sigma}, (\pi^{-1}\mathfrak{A})^{\Lambda}),$$

and the existence of the limit (6.3) follows:

$$h(\sigma, \mathfrak{A}) = h(\underline{\sigma}, \pi^{-1}\mathfrak{A}).$$

For any finite $\Lambda \subset \mathbf{Z}^{\nu}$, (6.4) gives

$$h(\underline{\sigma}, \pi^{-1}\mathfrak{A}) = h(\underline{\sigma}, (\pi^{-1}\mathfrak{A})^{\Lambda}).$$

* See Appendix A.3.2 and A.3.4.

Since $\text{diam}(\pi^{-1}\mathfrak{A})^\Lambda$ is arbitrarily small when diam $\mathfrak{A}$ is sufficiently small and Λ sufficiently large, Proposition 6.5(a) holds, and

$$h(\sigma) = h(\underline{\sigma}). \tag{6.20}$$

Concerning the definition of the pressure in Section 6.6, we note that (6.6) need not hold for a $\mathbf{Z}^\nu_\geqslant$-action, but that we have

$$Z_{\Lambda+x}(A, \mathfrak{A}) \leqslant Z_\Lambda(A, \mathfrak{A}) \quad \text{for} \quad \Lambda \subset \mathbf{Z}^\nu_\geqslant, \quad x \in \mathbf{Z}^\nu_\geqslant$$

because $\tau^x \Omega \subset \Omega$. Therefore $a \mapsto \log Z_{\Lambda(a)}(A, \mathfrak{A})$ remains subadditive, and (6.8), (6.10) hold. Furthermore

$$P(A, \mathfrak{A}) = \lim_{a\to\infty} |\Lambda(a)|^{-1} \log Z^*_{\Lambda(a)}(A, \mathfrak{A}),$$

where

$$Z^*_\Lambda(A, \mathfrak{A}) = \lim_{x\to\infty} Z_{\Lambda+x}(A, \mathfrak{A}).$$

Let $Z'_\Lambda(A, \mathfrak{A})$ denote the partition function computed from the restriction of A and $\mathfrak{A}$ to Ω' (defined by (6.19)). We have

$$Z'_\Lambda(A, \mathfrak{A}) = Z_\Lambda(A \circ \pi, \pi^{-1}\mathfrak{A})$$

and

$$Z'_\Lambda(A, \mathfrak{A}) \leqslant Z^*_\Lambda(A, \mathfrak{A}) \leqslant Z'_\Lambda(A, \mathfrak{A}) e^{|\Lambda|\delta},$$

where δ is the maximum of the oscillation of A on a set $\mathfrak{A}_i$. Also

$$\lim_{\text{diam}\mathfrak{A}\to 0} P(A \circ \pi, \pi^{-1}\mathfrak{A}) = P(A \circ \pi),$$

because $\text{diam}\,(\pi^{-1}\mathfrak{A})^\Lambda$ is arbitrarily small when diam $\mathfrak{A}$ is sufficiently small and Λ sufficiently large (see end of Section 6.6). Therefore finally (6.11) holds with

$$P(A) = P(A \circ \pi). \tag{6.21}$$

Sections 6.7, 6.8, 6.9 carry over to $\mathbf{Z}^\nu_\geqslant$-actions without any problem. Lemmas 6.10 and 6.11 will not be needed (but carry over). Theorem 6.12 remains correct because the case of a $\mathbf{Z}^\nu_\geqslant$-action reduces to that of a $\mathbf{Z}^\nu$-action by (6.20) and (6.21). Finally, with the definitions of $\boldsymbol{I}_A$, $\boldsymbol{I}'_A$ given in Section 6.13, Theorem 6.14 and Remark 6.15 remain correct.

6.19 Remark

If τ^* is a $\mathbf{Z}^\nu$-action on Ω, and τ the $\mathbf{Z}^\nu_\geqslant$-action obtained by restriction of τ^* to $\mathbf{Z}^\nu_\geqslant$, the τ-invariant states coincide with the τ^*-invariant states, $h_\tau = h_{\tau^*}$, and $P_\tau = P_{\tau^*}$. In fact the map $\pi : \underline{\Omega} \mapsto \Omega$ introduced in Section 6.17 is a homeomorphism.

6.20 Topological entropy

$P_\tau(0)$ is called the *topological entropy* of the $\mathbf{Z}^\nu_\geqslant$-action τ. From the various definitions of Sections 6.6 and 6.7 one sees that the topological entropy gives a measure of how mixing τ is. We have for instance

$$P(0) = \sup_{\mathfrak{A}} \inf_a |\Lambda(a)|^{-1} \log \text{ (least cardinal of a subcover of } \mathfrak{A}^{\Lambda(a)}).$$

The variational principle becomes here

$$P(0) = \sup_{\sigma \in \boldsymbol{I}} h(\sigma).$$

If $\nu > 1$ and $P_\tau(0) > 0$, it is easily seen that the topological entropy of the $\mathbf{Z}_\geqslant$-action associated with each generator of τ is $+\infty$.

6.21 Relative pressure

Let Ω and Ω' be metrizable compact sets with $\mathbf{Z}^\nu_\geqslant$ actions τ and τ' each generated by ν continuous maps. We assume that the diagram

$$\begin{array}{ccc} \Omega' & \xrightarrow{\tau'} & \Omega' \\ \downarrow \pi & & \downarrow \pi \\ \Omega & \xrightarrow{\tau} & \Omega \end{array}$$

is commutative, where π is continuous and onto Ω. We denote by $\boldsymbol{I}$ (respectively $\boldsymbol{I}'$) the set of τ-invariant probability measures on Ω (respectively of τ'-invariant probability measures on Ω').

Let $A' \in \mathcal{C}(\Omega')$, $\xi \in \Omega$, and Λ be a finite subset of $\mathbf{Z}^\nu_\geqslant$.

If $\mathfrak{A}$ is a finite open cover of Ω', we write

$$Z_\Lambda(A',\xi,\mathfrak{A}) = \min\left\{\sum_j \exp\left[\sup_{\xi'\in\mathfrak{B}_j\cap\pi^{-1}\xi}\sum_{x\in\Lambda}A'(\tau'^x\xi')\right]\right.$$

$$\left. : (\mathfrak{B}_j) \text{ is a subfamily of } \mathfrak{A}^\Lambda \text{ which covers } \pi^{-1}\xi\right\},$$

and define $Z^{(1)}_\Lambda$ similarly with $\inf_{\xi'}$ replacing $\sup_{\xi'}$.

If $\varepsilon > 0$ we write*

$$Z^{(2)}_\Lambda(A',\xi,\varepsilon) = \sup\left\{\sum_{\xi\in S}\exp\sum_{x\in\Lambda}A'(\tau'^x\xi')\right.$$

$$\left. : S\subset\pi^{-1}\xi \text{ and } S \text{ is } (\Lambda,\varepsilon)\text{-separated}\right\},$$

$$Z^{(3)}_\Lambda(A',\xi,\varepsilon) = \inf\left\{\sum_{\xi\in S}\exp\sum_{x\in\Lambda}A'(\tau'^x\xi')\right.$$

$$\left. : S\subset\Omega \text{ and } S \text{ is } (\Lambda,\varepsilon)\text{-spanning for } \pi^{-1}\xi\right\}.$$

With this notation the following results hold.

6.22 Theorem†

(a) *If $\sigma\in I$, for σ-almost all ξ the following limit exists:*

$$\begin{aligned}
P(A',\xi) &= \lim_{\operatorname{diam}\mathfrak{A}\to 0}\limsup_{a\to\infty}\frac{1}{|\Lambda(a)|}\log Z_{\Lambda(a)}(A',\xi,\mathfrak{A})\\
&= \lim_{\operatorname{diam}\mathfrak{A}\to 0}\limsup_{a\to\infty}\frac{1}{|\Lambda(a)|}\log Z^{(1)}_{\Lambda(a)}(A',\xi,\mathfrak{A})\\
&= \lim_{\varepsilon\to 0}\limsup_{a\to\infty}\frac{1}{|\Lambda(a)|}\log Z^{(2)}_{\Lambda(a)}(A',\xi,\varepsilon)\\
&= \lim_{\varepsilon\to 0}\limsup_{a\to\infty}\frac{1}{|\Lambda(a)|}\log Z^{(3)}_{\Lambda(a)}(A',\xi,\varepsilon),
\end{aligned}$$

and it defines a measurable τ-invariant function $P(A',\cdot)$ on Ω.

* Other possible definitions would be to take $S\subset\pi^{-1}\xi$ maximal (Λ,ε)-separated in $Z^{(2)}$ (instead of sup) and $S\subset\pi^{-1}\xi$ in $Z^{(3)}$ (instead of $S\subset\Omega$).

† Ledrappier and Walters [1].

(b) *The following variational principle holds*:

$$\int P(A', \xi)\sigma(d\xi) = \sup_{\sigma' \in I' : \pi\sigma' = \sigma} \left[h_{\tau'}(\sigma'|\pi) + \sigma'(A')\right], \qquad (6.22)$$

where the relative entropy $h_{\tau'}(\sigma'|\pi)$ *reduces to* $h_{\tau'}(\sigma') - h_{\tau}(\pi\sigma')$ *when* $h\tau(\pi\sigma') < +\infty$.

If σ is ergodic one can of course assume that $P(A', \xi)$ is independent of ξ, and write $P(A', \sigma)$ instead.

6.23 Corollary

Let $\nu = 1$, *F be a finite set, and* $M : \Omega \mapsto \mathbf{R}^{F^2}$ *be a continuous function with value in the* $F \times F$ *matrices with entries* $M_{jk} > 0$. *The limit*

$$P_\xi = \lim_{n\to\infty} \frac{1}{n} \log \|M(\xi)M(\tau\xi)\cdots M(\tau^{n-1}\xi)\| \qquad (6.23)$$

exists for σ*-almost all* ξ. *If we take* $\Omega' = \Omega \times F^{\mathbf{Z}}$ *and let* $\pi : \Omega' \mapsto \Omega$ *be the canonical projection, we have* $P_\xi = P(A', \xi)$, *where*

$$A'(\xi, (\xi_x)_{x\in\mathbf{Z}}) = \log M(\xi)_{\xi_0\xi_1},$$

and the formula (6.22) *holds.*

[For the existence of the limit (6.23) see Furstenberg and Kesten [1], Oseledec [1].]

Notes

The notion of entropy is due to Kolmogorov and Sinai. It is now classical for $\nu = 1$ (see Billingsley [1]). For $\nu > 1$, see Conze [1].

Much of the rest of the theory of this chapter is due to Walters [1]. He defined topological pressure and proved the variational principle (Theorem 6.12) for $\nu = 1$ (actually for a continuous map); the extension to $\nu > 1$ was done by Elsanousi [1].

Walters's variational principle is a generalization of the variational principle for lattice systems (Theorem 3.12). Walters was inspired in part by an intermediate generalization (Ruelle [4], where the "pressure" is introduced), and in part by the earlier work on topological entropy (see Section 6.20). Topological entropy was first defined by Adler, Konheim, and McAndrew [1]; equivalent definitions were later given by Bowen [3]. Goodwyn [1] proved that $P(0) \geqslant h(\sigma)$

for all $\sigma \in \boldsymbol{I}$. That $P(0)$ is the supremum of the $h(\sigma)$ was shown by Dinaburg [1] for finite-dimensional $\boldsymbol{\Omega}$, and then by Goodman [1] in general. Technically the idea contained in Lemma 6.10, and due to Goodwyn, played an essential role.

In the presentation given here we have largely followed Bowen [6], except for the simple proof of one half of Theorem 6.12, due to Denker [1]. Another very simple proof of the whole theorem is due to Misiurewicz [1]. Theorem 6.14 in this setting appears to be new. In general, the reader is urged to consult Bowen [6] for further details.

Exercises

1. [See Bowen [6], Lemma 2.2 and Lemma 2.3.]
   (a) Let $\mathfrak{A}', \mathfrak{A}'', \mathfrak{B}$ be Borel partitions of $\boldsymbol{\Omega}$. Show that

$$H(\sigma, \mathfrak{A}' \vee \mathfrak{A}'' \vee \mathfrak{B}) - H(\sigma, \mathfrak{A}' \vee \mathfrak{B}) - H(\sigma, \mathfrak{A}'' \vee \mathfrak{B}) + H(\sigma, \mathfrak{B}) \leqslant 0.$$

[This is, in a new guise, the strong subadditivity of the entropy; see Equation (3.22).] This can be rewritten

$$\begin{aligned} &H(\sigma, \mathfrak{A}' \vee \mathfrak{A}'' \vee \mathfrak{B}) - H(\sigma, \mathfrak{B}) \\ &\quad \leqslant H(\sigma, \mathfrak{A}' \vee \mathfrak{B}) - H(\sigma, \mathfrak{B}) + H(\sigma, \mathfrak{A}'' \vee \mathfrak{B}) - H(\sigma, \mathfrak{B}). \end{aligned}$$

Therefore if $\mathfrak{A}, \mathfrak{B}$ are Borel partitions we have

$$\begin{aligned} H(\sigma, \mathfrak{A}^\Lambda) - H(\sigma, \mathfrak{B}^\Lambda) &\leqslant H(\sigma, \mathfrak{A}^\Lambda \vee \mathfrak{B}^\Lambda) - H(\sigma, \mathfrak{B}^\Lambda) \\ &\leqslant \sum_{x \in \Lambda} [H(\sigma, (\tau^{-x}\mathfrak{A}) \vee \mathfrak{B}^\Lambda) - H(\sigma, \mathfrak{B}^\Lambda)] \\ &\leqslant \sum_{x \in \Lambda} [H(\sigma, (\tau^{-x}\mathfrak{A}) \vee (\tau^{-x}\mathfrak{B})) - H(\sigma, \tau^{-x}\mathfrak{B})] \\ &= |\Lambda| [H(\sigma, \mathfrak{A} \vee \mathfrak{B}) - H(\sigma, \mathfrak{B})]. \end{aligned}$$

From this (6.5) follows.

   (b) Given $\mathfrak{A}$, σ, and $\varepsilon > 0$, show that one can choose $\delta > 0$ such that

$$H(\sigma, \mathfrak{A} \vee \mathfrak{B}) - H(\sigma, \mathfrak{B}) < \varepsilon$$

if diam $\mathfrak{B} < \delta$.

[Denote by Δ the symmetric difference: $A \Delta B = (A \cup B) \setminus (A \cap B)$. One can choose $\alpha > 0$ such that if the partition $\mathfrak{A}'$ has the same index set as $\mathfrak{A}$ and

$$\sigma(\mathfrak{A}_i \Delta \mathfrak{A}'_i) < \alpha \qquad \text{for all } i, \tag{*}$$

then

$$H(\sigma, \mathfrak{A} \vee \mathfrak{A}') - H(\sigma, \mathfrak{A}') = -\sum_{i,k} \sigma(\mathfrak{A}_i \cap \mathfrak{A}'_k) \log \frac{\sigma(\mathfrak{A}_i \cap \mathfrak{A}'_k)}{\sigma(\mathfrak{A}'_k)} < \varepsilon.$$

For diam $\mathfrak{B} < \delta$ sufficiently small there exists a partition $\mathfrak{A}'$ coarser than $\mathfrak{B}$ (i.e. its sets are unions of sets of $\mathfrak{B}$) such that (*) holds; then

$$\begin{aligned} H(\sigma, \mathfrak{A} \vee \mathfrak{B}) - H(\sigma, \mathfrak{B}) &= H(\sigma, \mathfrak{A} \vee \mathfrak{A}' \vee \mathfrak{B}) - H(\sigma, \mathfrak{A}' \vee \mathfrak{B}) \\ &\leqslant H(\sigma, \mathfrak{A} \vee \mathfrak{A}') - H(\sigma, \mathfrak{A}') < \varepsilon \big]. \end{aligned}$$

2. We have defined Z_Λ (Section 6.6), and $Z_\Lambda^{(i)}$ for $i = 1, 2, 3$ (Section 6.7). Let also

$$Z_\Lambda^{(2)'}(A, \varepsilon) = \sum_{\xi \in S} \exp \sum_{x \in \Lambda} A(\tau^x \xi),$$

where S is an arbitrarily chosen maximal (Λ, ε)-separated set.
(a) $Z_\Lambda^{(3)} \leqslant Z_\Lambda^{(2)'} \leqslant Z_\Lambda^{(2)}$ [a maximal (Λ, ε)-separated set is (Λ, ε)-spanning].
(b) $Z_\Lambda^{(2)}(A, \operatorname{diam} \mathfrak{A}) \leqslant Z_\Lambda(A, \mathfrak{A})$.
(c) $Z_\Lambda^{(1)}(A, \mathfrak{A}) \leqslant Z_\Lambda^{(3)}(A, \delta/2)$ if δ is a Lebesgue number for $\mathfrak{A}$.
(d) Prove the assertions of Section 6.7.
(e) Define P in terms of $Z_\Lambda^{(2)'}$.

3. Let $\mathbf{\Omega}, \mathbf{\Omega}^*$ be compact metrizable with $\mathbf{Z}_\geqslant^\nu$-actions τ and τ^*. Let $f : \mathbf{\Omega}^* \to \mathbf{\Omega}$ be continuous and such that $f \circ \tau^{*x} = \tau^x \circ f$.
(a) If σ^* is a τ^*-invariant probability measure on $\mathbf{\Omega}^*$, then $h_{\tau^*}(\sigma^*) \geqslant h_\tau(f\sigma^*)$. If f is injective we have equality.
(b) If A is a real continuous function on $\mathbf{\Omega}$ and f is surjective, then

$$P_\tau(A) \leqslant P_{\tau^*}(A \circ f).$$

(c) If A is a real continuous function on $\mathbf{\Omega}$ and f is injective, then

$$P_\tau(A) \geqslant P_{\tau^*}(A \circ f)$$

($P_\tau^*(A \circ f)$ is the pressure of A restricted to $f\mathbf{\Omega}^*$). If the support of every τ-invariant measure is contained in $f\mathbf{\Omega}^*$, we have equality.

4. Suppose that $\mathbf{\Omega} = \cup_\alpha \mathbf{\Omega}^{(\alpha)}$, where the $\mathbf{\Omega}^{(\alpha)}$ are closed and τ-stable (i.e. $\tau^x \mathbf{\Omega}^{(\alpha)} \subset \mathbf{\Omega}^{(\alpha)}$ for all $x \in \mathbf{Z}_\geqslant^\nu$). Then $P_\tau(A)$ is the sup over α of the pressure of A restricted to $\mathbf{\Omega}^{(\alpha)}$. [Choose $\sigma \in \boldsymbol{I}$ such that $P(A) < h(\sigma) + \sigma(A) + \varepsilon$, and decompose σ into measures with appropriate supports; cf. Walters [1], Corollary 4.12(i).]

5. Let $\mathbf{\Omega}_1, \mathbf{\Omega}_2$ be compact metrizable with $\mathbf{Z}_\geqslant^\nu$ actions τ_1, τ_2, and let $\tau^x = \tau_1^x \times \tau_2^x$ on $\mathbf{\Omega}_1 \times \mathbf{\Omega}_2$. If $A_1 \in \mathcal{C}(\mathbf{\Omega}_1)$, $A_2 \in \mathcal{C}(\mathbf{\Omega}_2)$, and $A(\xi_1, \xi_2) = A(\xi_1) + A(\xi_2)$, then $P_\tau(A) = P_{\tau_1}(A_1) + P_{\tau_2}(A_2)$. [Cf. Walters [1], Theorem 2.2(viii).

Use the definition of P in terms of $Z^{(3)}$ to prove that $P_\tau(A) \leqslant P_{\tau_1}(A_1) + P_{\tau_2}(A_2)$, and the variational principle to prove the reverse inequality.]

6. If $P(0) < +\infty$, then

$$\{\sigma \in \mathcal{C}^* : \sigma(A) \leqslant P(A) \text{ for all } A \in \mathcal{C}\} = \boldsymbol{I}.$$

7. If $P(0) < +\infty$, let $\boldsymbol{I}'_A$ be defined by (6.17). Show that $\boldsymbol{I}_A \subset \boldsymbol{I}'_A$, and that the set $\boldsymbol{D}' = \{A \in \mathcal{C} : \boldsymbol{I}'_A \text{ is reduced to a point}\}$ is residual.

Show that $\boldsymbol{I}'_A$ is equal to the set $\boldsymbol{I}''_A$ of limits $\boldsymbol{\sigma}$ of sequences $\boldsymbol{\sigma}_n$ such that $\lim_{n\to\infty}[h(\boldsymbol{\sigma}_n) + \boldsymbol{\sigma}_n(A)] = P(A)$.

[Let $\boldsymbol{\sigma} = \lim \boldsymbol{\sigma}_n$ where $h(\boldsymbol{\sigma}_n) + \boldsymbol{\sigma}_n(A) \geqslant P(A) - 1/n$. Since $P(A+B) \geqslant h(\boldsymbol{\sigma}_n) + \boldsymbol{\sigma}_n(A+B) \geqslant P(A) + \boldsymbol{\sigma}_n(B) - 1/n$ for all $B \in \mathcal{C}$, we have $\boldsymbol{\sigma} \in \boldsymbol{I}'_A$, and hence $\boldsymbol{I}''_A \subset \boldsymbol{I}'_A$. Therefore if $A \in \boldsymbol{D}'$, then $\boldsymbol{I}''_A = \boldsymbol{I}'_A$. To show that $\boldsymbol{I}'_A \subset \boldsymbol{I}''_A$, use the fact that $\boldsymbol{I}'_A$ is contained in the closed convex hull of limits of $\boldsymbol{\sigma}_n$, where $\{\boldsymbol{\sigma}_n\} = \boldsymbol{I}'_{A_n}$, $A_n \in \boldsymbol{D}$, and $A_n \to A$ (Appendix A.3.7).]

7

Statistical mechanics on Smale spaces

Part of the thermodynamic formalism can be extended to the case of a general $\mathbf{Z}^\nu$-action by homeomorphisms of a compact metrizable space $\boldsymbol{\Omega}$, as we have seen in Chapter 6. We shall here extend the richer formalism of one-dimensional systems, described in Chapter 5, to a certain class of $\mathbf{Z}$-actions by homeomorphisms of compact metric spaces. Such $\mathbf{Z}$-actions have been met and studied in the theory of axiom A diffeomorphisms by Smale [1]. We present here an abstract version of the relevant part of this theory. For the proofs we shall mostly refer to the publications on axiom A diffeomorphisms. These publications should also be consulted for motivation, particularly Smale [1] and Bowen [6]. The main new assumption here is that of a *local product structure*. The space $\boldsymbol{\Omega}$ is "foliated" by "stable manifolds" V_x^+ which contract exponentially under iterates of f, and "unstable manifolds" V_x^- which contract under iterates of f^{-1}. If x and y are sufficiently close, V_x^+ and V_y^- determine a point $[x, y] \in V_x^+ \cap V_y^-$ and the local product structure is defined by the map $x, y \to [x, y]$.

7.1 Smale spaces

Let $\boldsymbol{\Omega}$ be a non-empty compact metric space with metric d. We assume that we are given $\varepsilon > 0$ and $[\cdot, \cdot]$ with the following property:
(SS1)

$$[\cdot, \cdot] : \{(x, y) \in \boldsymbol{\Omega} \times \boldsymbol{\Omega} : d(x, y) < \varepsilon\} \to \boldsymbol{\Omega}$$

is a continuous map such that $[x,x] = x$ *and*

$$[[x, y], z] = [x, z], \qquad [x, [y, z]] = [x, z]$$

when the two sides of these relations are defined.

Note that there exists $\varepsilon_n > 0$ such that any expression with nested square brackets of points $x_1, \ldots, x_n$ (in this order) is defined and equal to $[x_1, x_n]$ when diam $\{x_1, \ldots, x_n\} < \varepsilon_n$.

Let in particular $d(x, y) < \varepsilon_4$. If we write $u = [y, x]$, $v = [x, y]$, then

$$\begin{array}{lll} d(u, x) < \varepsilon, & d(v, x) < \varepsilon, & d(u, v) < \varepsilon, \\ u = [u, x], & v = [x, v], & [u, v] = y. \end{array} \tag{7.1}$$

Conversely, if u, v satisfy these conditions, then $u = [u, x] = [[u, v], x] = [y, x]$ and similarly $v = [x, y]$. This implies that $[\cdot, \cdot] : V_x^-(\delta) \times V_x^+(\delta) \mapsto \mathbf{\Omega}$ is a homeomorphism onto an open subset of $\mathbf{\Omega}$, where we have defined

$$V_x^-(\delta) = \{u : u = [u, x] \quad \text{and} \quad d(x, u) < \delta\}, \tag{7.2}$$

$$V_x^+(\delta) = \{v : v = [x, v] \quad \text{and} \quad d(x, v) < \delta\}, \tag{7.3}$$

and chosen δ such that $d(x, u) < \delta$ and $d(x, v) < \delta$ imply $d(x, [u, v]) < \varepsilon_4$.

If $d(x, y) < \varepsilon_4$, we obtain from (7.1), (7.2), (7.3)

$$\{[x, y]\} = V_x^+(\varepsilon) \cap V_y^-(\varepsilon). \tag{7.4}$$

Assume now given $\lambda \in (0, 1)$ and f with the following property:

(SS2) *f is a homeomorphism of $\mathbf{\Omega}$ such that $f[x, y] = [fx, fy]$ when both sides are defined and*

$$\begin{array}{llll} d\,(f^n y, f^n z) \leqslant \lambda^n d\,(y, z) & \text{if} & y, z \in V_x^+(\delta), & n > 0, \\ d\,(f^{-n} y, f^{-n} z) \leqslant \lambda^n d\,(y, z) & \text{if} & y, z \in V_x^-(\delta), & n > 0. \end{array}$$

Replacing possibly δ by a smaller number, we have

$$V_x^+(\delta) = \{y : d(f^n x, f^n y) < \delta \quad \text{for all } n \geqslant 0\}, \tag{7.5}$$

$$V_x^-(\delta) = \{y : d(f^{-n} x, f^{-n} y) < \delta \quad \text{for all } n \geqslant 0\}. \tag{7.6}$$

[Choose δ' such that $d(x, y) < \delta'$ implies

$$d(x, [x, y]) < \delta, d(x, [y, x]) < \delta.$$

We have the following implications of properties each valid for all $n \geqslant 0$:

$$\begin{aligned} & d(f^n x, f^n y) < \delta' \Rightarrow d(f^n x, f^n [y, x]) < \delta \\ & \quad \Rightarrow f^n [y, x] \in V_{f^n x}^-(\delta) \\ & \quad \Rightarrow d(x, [y, x]) \leqslant \lambda^n d(f^n x, f^n [x, y]) \leqslant \lambda^n \delta \\ & \quad \Rightarrow d(x, [y, x]) = 0 \Rightarrow y = [[y, x], [x, y]] = [x, [x, y]] = [x, y]. \end{aligned}$$

Therefore

$$V_x^+(\delta') \supset \{y : d(f^n x, f^n y) < \delta' \text{ for all } n \geqslant 0\}.$$

The converse inclusion follows from (SS2). We have thus proved (7.5) with δ replaced by the smaller number δ'. The proof of (7.6) is analogous].

We define a *Smale space* to be a compact metric space $\boldsymbol{\Omega}$ with a map $[\cdot, \cdot]$ and a homeomorphism f satisfying (SS1) and (SS2) for suitable ε, λ.

Notice that there is a natural *duality* which replaces f by f^{-1}, $[x, y]$ by $[y, x]$, V^+ by V^-, etc.

7.2 Example

Let Ω be the configuration space of a $\mathbf{Z}$-lattice system (Ω_0, t), and τ the corresponding shift (see Chapter 5). Given $\lambda \in (0, 1)$, we define a distance d on Ω by

$$d(\xi, \eta) = \lambda^k \tag{7.7}$$

if $\xi = (\xi_n)_{n\in\mathbf{Z}} \in \Omega$, $\eta = (\eta_n)_{n\in\mathbf{Z}} \in \Omega$, and $k = \inf\{|n| : \xi_n \neq \eta_n\}$.

If $d(\xi, \eta) < 1$, then $\xi_0 = \eta_0$ and we can define

$$[\xi, \eta] = (\ldots, \eta_{-l}, \ldots, \eta_{-1}, \xi_0, \xi_1, \ldots, \xi_l, \ldots) \in \Omega.$$

It is then easy to check that the conditions of Section 7.1 are satisfied with $\boldsymbol{\Omega} = \Omega$, $f = \tau$, $\varepsilon = \delta = 1$.

It is also readily seen that the Banach space $\mathcal{C}^\alpha(\boldsymbol{\Omega})$ of real Hölder continuous functions of exponent α is identical with the space $\mathcal{F}^\theta$ introduced in Section 5.19, with $\theta = \lambda^{\alpha/2}$.

7.3 Properties of Smale spaces

f is expansive (see Section 6.2) with expansive constant δ if δ is sufficiently small [obvious from (7.5), (7.6), (7.4)]. More precisely, there is $C > 0$ such that if $d(f^k x, f^k y) < \delta$ for $|k| < n$, then $d(x, y) < C\lambda^n$. [We can take $C = \max$ (diam $\boldsymbol{\Omega}$, $2\varepsilon/\lambda$). Using (SS2) we have indeed for $n \neq 0$

$$d(x, [x, y]) \leqslant \lambda^{n-1} d(f^{-n+1}x, f^{-n+1}[x, y]),$$
$$d([x, y], y) \leqslant \lambda^{n-1} d(f^{n-1}[x, y], f^{n-1}y).]$$

Let S be a finite or infinite interval* of $\mathbf{Z}$, $\boldsymbol{x} = (x_k)_{k\in S} \in \boldsymbol{\Omega}^S$, and $\alpha > 0$. We say that $\boldsymbol{x}$ is an *α-pseudo-orbit* if

$$d(fx_k, x_{k+1}) < \alpha \qquad \text{whenever} \quad k, k+1 \in S.$$

We say that $x \in \boldsymbol{\Omega}$ *α-shadows* $\boldsymbol{x}$ if

$$d(f^k x, x_k) < \alpha \qquad \text{for all} \quad k \in S.$$

Given $\beta > 0$ there is $\alpha > 0$ so that

* S is a finite interval $[k, l]$ (see Section 5.17) or $[k, +\infty)$, or $(-\infty, l]$, or $\mathbf{Z}$.

(a) *every α-pseudo-orbit x is β-shadowed by a point $x \in \Omega$,*
(b) *if $x \in \Omega$ and $d(f^n x, x) < \alpha$, then there is $y \in \Omega$ with $f^n y = y$ and*

$$d(f^k x, f^k y) < \beta \qquad \text{for all} \quad k \ni [0, n].$$

[See Bowen [6], Proposition 3.6 and Corollary 3.7.]

The non-wandering set of (Ω, f),

$$\left\{x \in \Omega : U \cap \bigcup_{n>0} f^n U \neq \emptyset \text{ for every open } U \ni x\right\},$$

is the closure of the set

$$\{x \in \Omega : f^n x = x \text{ for some } n > 0\}$$

of periodic points.

This is *Anosov's closing lemma* [see Bowen [6], 3.8.] Notice that the non-wandering set is not empty (see Appendix A.2).

7.4 Smale's "spectral decomposition"

*The non-wandering set of (Ω, f) is the union of finitely many disjoint compact sets Ω^α, with $f\Omega^\alpha = \Omega^\alpha$, and $f|\Omega^\alpha$ topologically + transitive.**

Each Ω^α is the union of n_α disjoint compact sets $\Omega^{\alpha\beta}$ which are cyclically permuted by f and such that $f^{n_\alpha}|\Omega^{\alpha\beta}$ is topologically mixing.

These properties determine uniquely the Ω^α, n_α, and $\Omega^{\alpha\beta}$.

[See Bowen [6], Theorem 3.5.] This extends part of Theorems 5.2 and 5.3 on **Z**-lattice systems. Notice that the non-wandering set of (Ω, f) and the sets Ω^α are again Smale spaces; $\Omega^{\alpha\beta}$ is a Smale space with respect to $f^{n_\alpha}|\Omega^{\alpha\beta}$.

The Ω^α are called *basic sets*. Any f-invariant measure on Ω has its support in the non-wandering set. In particular, every f-ergodic probability measure has its support in one of the basic sets.

7.5 Markov partitions and symbolic dynamics

Let $\delta > 0$ be sufficiently small and $x \in \Omega$. Suppose that $C \subset V_x^-(\delta)$ is the closure of its interior in $V_x^-(2\delta)$, and $D \subset V_x^+(\delta)$ the closure of its interior in

* See Appendix A.2 for the definition of topological + transitivity, transitivity, and mixing.

$V_x^+(2\delta)$. Then $R = [C, D]$ is the closure of its interior in $\mathbf{\Omega}$, and is called a *rectangle*. Its boundary is

$$\partial R = \partial^+ R \cup \partial^- R,$$

where

$$\partial^+ R = [\partial C, D], \qquad \partial^- R = [C, \partial D],$$

and ∂C, ∂D are the boundaries of C, D in $V_x^-(2\delta)$, $V_x^+(2\delta)$ respectively.

A *Markov partition* is a finite cover (R_i) of $\mathbf{\Omega}$ by rectangles such that

(a) $\text{int}\, R_i \cap \text{int} R_j = \emptyset$ if $i \neq j$;
(b) if $x \in \text{int}\, R_i \cap f^{-1}\, \text{int}\, R_j$, then

$$f[C_i, x] \supset [C_j, fx],$$
$$f[x, D_i] \subset [fx, D_j],$$

where we have written $R_i = [C_i, D_i]$, $R_j = [C_j, D_j]$.

A Smale space $\mathbf{\Omega}$ *has Markov partitions of arbitrarily small diameter.*

[See Bowen [1]; or [6], Theorem 3.12.]

If (R_i) is a Markov partition, we write

$$\partial^+ = \bigcup_i \partial^+ R_i, \qquad \partial^- = \bigcup_i \partial^- R_i.$$

One can then show that

$$f\partial^+ \subset \partial^+, \qquad f^{-1}\partial^- \subset \partial^-.$$

Let Ω_0 be the set of rectangles R_i in a Markov partition, and define

$$t_{R_i R_j} = \begin{cases} 1 & \text{if} \quad \text{int} R_i \cap f^{-1} \text{int} R_j \neq \emptyset, \\ 0 & \text{otherwise.} \end{cases}$$

The set Ω_0 (*symbol set*) and the matrix t (*transition matrix*) define a $\mathbf{Z}$-lattice system (Ω_0, t) (see Chapter 5), with configuration space Ω and shift τ. The dynamical system (Ω, τ) constitutes the *symbolic dynamics* of the dynamical system $(\mathbf{\Omega}, f)$. This terminology is justified by the following results, valid for a Markov partition of sufficiently small diameter.

7.6 Theorem

If $\xi = (\xi_n)_{n \in \mathbf{Z}} \in \Omega$, *then* $\bigcap_{n \in \mathbf{Z}} f^{-n} \xi_n$ *consists of a single point* $\pi(\xi)$. *Furthermore*

(a) *The map* $\pi : \Omega \mapsto \mathbf{\Omega}$ *is continuous onto.*
(b) $\pi \circ \tau = f \circ \pi$.
(c) π^{-1} *is uniquely defined on the residual set* $\mathbf{\Omega} \setminus \cup_{n\in\mathbf{Z}} f^n(\partial^+ \cup \partial^-)$.
(d) *There is an integer d such that, for all* x, $\pi^{-1}x$ *has at most d elements.*
(e) *If f is topologically + transitive (respectively mixing), then* (Ω_0, t) *is transitive (respectively mixing).*

For (a), (b), (c), (e) see Bowen [1], Section 4; or [6], Theorem 3.18 and Proposition 3.19. For (d) see Bowen [2], Proposition 10; one can in fact take $d = |\Omega_0|^2$ (R. Bowen, private communication).

7.7 Hölder continuous functions

Part (a) of the above theorem can be made more precise. *If the number* λ *used in* (7.7) *to define the metric on* Ω *is the same as appears in* (SS2) *for* $\mathbf{\Omega}$, *then* π *is Lipshitz, i.e., there exists* $C > 0$ *such that*

$$d(\pi\xi, \pi\eta) \leqslant Cd(\xi, \eta). \tag{7.8}$$

If $\xi \neq \eta$ we have $d(\xi, \eta) = \lambda^n$ for some $n \geqslant 0$, and $\xi_k = \eta_k$ for $|k| < n$. Assuming that the diameter of the Markov partition is $< \delta$, we have

$$d(f^k\pi\xi, f^k\pi\eta) < \delta \qquad \text{for} \quad |k| < n,$$

and therefore

$$d(\pi\xi, \pi\eta) < C\lambda^n,$$

as noted in Section 7.3. From this (7.8) follows.

Since π is Lipshitz, if $A \in \mathcal{C}^\alpha(\mathbf{\Omega})$, i.e., if A is a Hölder continuous function of exponent α on $\mathbf{\Omega}$, then $A \circ \pi \in \mathcal{C}^\alpha(\Omega) = \mathcal{F}^\theta$ with $\theta = \lambda^{\alpha/2}$ (see Section 7.2). This defines a bounded linear map $\mathcal{C}^\alpha(\mathbf{\Omega}) \mapsto \mathcal{F}^\theta$.

7.8 Pressure and equilibrium states

The study of pressure and equilibrium states can be reduced to the same problem for basic sets (see Section 7.4 and Exercises 3(c) and 4 of Chapter 6). We may thus suppose that $(\mathbf{\Omega}, f)$ is topologically + transitive. Notice that

$$P_f(A) \leqslant P_\tau(A \circ \pi) \tag{7.9}$$

(see Exercise 3(b) of Chapter 6).

Suppose now that $A \in \mathcal{C}^\alpha(\boldsymbol{\Omega})$, with $(\boldsymbol{\Omega}, f)$ topologically + transitive. Since $A \circ \pi \in \mathcal{F}^\theta$ with $\theta = \lambda^{\alpha/2}$, there is a unique τ-invariant probability measure ρ on Ω such that

$$h_\tau(\rho) + \rho(A \circ \pi) = P_\tau(A \circ \pi);$$

furthermore supp $\rho = \Omega$ (see Corollary 5.6(a)). The sets $\pi^{-1} \bigcap_{n \geqslant 0} f^n \partial^+, \pi^{-1} \bigcap_{n \geqslant 0} f^{-n} \partial^-$ are closed the τ-invariant. Their complements are not empty, and thus have ρ-measure different from zero. Using ergodicity gives

$$\rho\left(\pi^{-1} \bigcap_{n \geqslant 0} f^n \partial^+\right) = \rho\left(\pi^{-1} \bigcap_{n \geqslant 0} f^{-n} \partial^-\right) = 0;$$

hence

$$\rho(\pi^{-1}\partial^+) = \rho(\pi^{-1}\partial^-) = 0;$$

hence

$$\rho\left(\pi^{-1} \bigcup_{n \in \mathbf{Z}} f^n(\partial^+ \cup \partial^-)\right) = 0.$$

In view of Theorem 7.6(c), $\pi : (\Omega, \rho) \mapsto (\boldsymbol{\Omega}, \pi\rho)$ is an isomorphism of abstract dynamical systems. In particular

$$h_f(\pi\rho) = h_\tau(\rho),$$

and therefore

$$P_f(A) \geqslant h_f(\pi\rho) + (\pi\rho)(A) = P_\tau(A \circ \pi).$$

Comparing with (7.9) shows that

$$P_f(A) = P_\tau(A \circ \pi) \tag{7.10}$$

and that $\pi\rho$ is an equilibrium state for A.

If $\boldsymbol{\sigma}$ is any equilibrium state for A, there exists a τ-invariant state σ such that $\pi\sigma = \boldsymbol{\sigma}$ [use the Hahn–Banach and Markov–Kakutani theorems, Appendix A.3.2 and A.3.4]. Then

$$\begin{aligned} P_\tau(A \circ \pi) &\geqslant h_\tau(\sigma) + \sigma(A \circ \pi) \\ &\geqslant h_f(\boldsymbol{\sigma}) + \boldsymbol{\sigma}(A) = P_f(A) \end{aligned}$$

[use Exercise 3(a) of Chapter 6]. Thus σ is an equilibrium state for $A \circ \pi$; hence $\sigma = \rho$.

Notice that (7.10) remains true for $A \in \mathcal{C}(\boldsymbol{\Omega})$, because of the density of $\mathcal{C}^\alpha(\boldsymbol{\Omega})$ in $\mathcal{C}(\boldsymbol{\Omega})$. We have thus proved the following result.

7.9 Theorem

If $(\mathbf{\Omega}, f)$ is topologically + transitive and $A \in \mathcal{C}(\mathbf{\Omega})$, then

$$P_f(A) = P_\tau(A \circ \pi).$$

If $A \in \mathcal{C}^\alpha(\mathbf{\Omega})$, there is a unique equilibrium state $\rho_A = \pi\rho$ where ρ is the unique equilibrium state for $A \circ \pi$. The map $\pi : (\Omega, \rho) \to (\mathbf{\Omega}, \rho_A)$ is an isomorphism of abstract dynamical systems.

Here are a few consequences.

7.10 Corollary

Assume that $(\mathbf{\Omega}, f)$ is topologically + transitive.

(a) *The function P_f is real analytic on $\mathcal{C}^\alpha(\mathbf{\Omega})$.*
(b) *If $A \in \mathcal{C}^\alpha(\mathbf{\Omega})$, then supp $\rho_A = \mathbf{\Omega}$.*
(c) *Let $A, A' \in \mathcal{C}^\alpha(\mathbf{\Omega})$. Then $\rho_A = \rho_{A'}$ if and only if there exist $c \in R$ and C continuous such that*

$$A' - A = c + C \circ f - C.$$

c is unique, and C is unique up to an additive constant.
(d) *If $A \in \mathcal{C}^\alpha(\mathbf{\Omega})$ and $(\mathbf{\Omega}, f)$ is topologically mixing, then (ρ_A, f) is isomorphic to a Bernoulli shift.*

(a) follows from Corollary 5.27; (b) follows from Corollary 5.6(a); (c) is proved like Theorem 5.7 (use (a) and see Exercise 2); (d) follows from Theorem 5.10.

7.11 Remark

The function C in Corollary 7.10(c) is Hölder continuous: $C \in \mathcal{C}^\alpha(\mathbf{\Omega})$ when the following condition is satisfied (see Exercises 2, 3).

(SS3) *There is a constant $L > 0$ such that*

$$d(x, [x, y]) \leqslant Ld(x, y).$$

This condition will be used only when explicitly stated. It is satisfied in the example of Section 7.2.

Here are other consequences of Theorem 7.9.

7.12 Corollary

We assume that (Ω, f) is topologically mixing. Let $\alpha \in (0, 1)$ be fixed, and define

$$D^l(B_1, \ldots, B_l) = \frac{\partial^l}{\partial s_1 \cdots \partial s_l} P\left(A + \sum_i s_i B_i\right)\Bigg|_{s_1=\cdots=s_l=0}$$

for $A, B_1, \ldots, B_l \in \mathcal{C}^\alpha(\Omega)$. In particular $D^1(B_1) = \rho_A(B_1)$.

(a) *Given $M > 0$ there exist $a, b > 0$ such that*

$$\left|\rho_A(B_1 \cdot (B_2 \circ f^k)) - \rho_A(B_1)\rho_A(B_2)\right| \leqslant e^{a-b|k|}\|B_1\|_\alpha \|B_2\|_\alpha$$

whenever $\|A\|_\alpha \leqslant M$($\| \ \|_\alpha$ denotes the norm in $\mathcal{C}^\alpha(\Omega)$).

(b) $D^2(B_1, B_2) = \sum_{k\in\mathbf{Z}} [\rho_A(B_1 \cdot (B_2 \circ f^k)) - \rho_A(B_1)\rho_A(B_2)]$.

(c) *$B_1 \mapsto D^2(B_1, B_1)$ is a positive semi-definite quadratic form on $\mathcal{C}^\alpha(\Omega)$. If (SS3) holds, its kernel is $\{c + C \circ f - C : c \in R, C \in \mathcal{C}^\alpha(\Omega)\}$. There is $R > 0$ such that*

$$[D^2(B_1, B_1)]^{1/2} \leqslant R\|B_1\|_\alpha.$$

(d) *For all $p \in \mathbf{R}$ (mod 2π),*

$$\sum_{k\in\mathbf{Z}} e^{-ipk}[\rho_A(B_1 \cdot (B_1 \circ f^k)) - \rho_A(B_1)\rho_A(B_1)] \geqslant 0.$$

(e) (Central limit theorem.*) *Define*

$$B_\Lambda = |\Lambda|^{-1/2} \sum_{k\in\Lambda} [B_1 \circ f^k - \sigma(B_1)],$$

where Λ is a finite interval of $\mathbf{Z}$. Let γ be the gaussian probability measure $(1/\sqrt{2\pi D})e^{-t^2/2D}\ dt$ on $\mathbf{R}$ if $D = D^2(B_1, B_1) > 0$, the unit mass δ_0 at 0 if $D_2(B_1, B_1) = 0$. When $|\Lambda| \to \infty$, the measure $B_\Lambda\sigma$ tends vaguely to γ.

All this is a translation of Exercises 4(e), 5, and 9 of Chapter 5.

7.13 Corollary

Let (Ω, f) be topologically + transitive and $A, B \in \mathcal{C}^\alpha(\Omega)$. Write

$$Z_{[a,b]} = \rho_A\left(\exp \sum_{k=a}^{b-1} B \circ f^k\right).$$

* See Ratner [1]; this result can be improved and generalized following Exercise 9 of Chapter 5.

Then

(a) $\lim_{b-a\to\infty} \frac{1}{b-a} \log Z_{[a,b]} = P(A+B) - P(A)$.

(b) $Z_{[a,b]}^{-1}(\exp \sum_{k=a}^{b-1} B \circ f^k) \cdot \rho_A$ *tends to* ρ_{A+B} *in the vague topology when* $a \to -\infty, b \to +\infty$.

See Proposition 4.4 and Remark 4.5 for the mixing situation. The + transitive case follows readily.

7.14 Equilibrium states for A not Hölder continuous

In this section we describe a result valid for dynamical systems somewhat more general than Smale spaces.

Let Ω again be a general compact metrizable space, and $f : \Omega \mapsto \Omega$ a homeomorphism. Choose a metric d on Ω. We say that f satisfies *specification* if, for each $\delta > 0$, there is $p(\delta) > 0$ such that the following holds:

if $\Lambda_1, \ldots, \Lambda_n$ *are finite intervals of* $\mathbf{Z}$ *contained in* $[a, b]$, *with* $d(\Lambda_i, \Lambda_j) \geqslant p(\delta)$ *when* $i \neq j$, *and* $x_1, \ldots, x_n \in \Omega$, *then there is* $x \in \Omega$ *such that*

$$f^{b-a+p(\delta)}x = x,$$

and

$$d(f^k x, f^k x_i) < \delta \qquad for \quad k \in \Lambda_i.$$

Let A be a real continuous function on Ω. We introduce the condition (S) *There exist* $\delta > 0$ *and* $K \geqslant 0$ *such that if* $d(f^k x, f^k y) < \delta$ *for* $k = 0, 1, \ldots, n$, *then*

$$\left| \sum_{k=0}^{n} A(f^k x) - \sum_{k=0}^{n} A(f^k y) \right| \leqslant K.$$

With this terminology, Bowen [5] has obtained the following uniqueness criterion:

If the homeomorphism $f : \Omega \mapsto \Omega$ *is expansive and satisfies specification, and if* $A \in \mathcal{C}(\Omega)$ *satisfies condition* (S), *then* A *has a unique equilibrium state.*

In the case of Smale spaces, mixing implies specification, as can be seen using symbolic dynamics (Theorem 7.6(e)). We have thus the following corollary:

If (Ω, f) *is a topologically + transitive Smale space, and if* $A \in \mathcal{C}(\Omega)$ *satisfies condition* (S), *then* A *has a unique equilibrium state.*

[The topologically + transitive case reduces to the mixing case.]
For any $A \in \mathcal{C}(\Omega)$, define

$$\mathrm{var}_{n,\delta} A = \sup\{|A(y) - A(x)| : d(f^k x, f^k y) < \delta \text{ for } |k| \leqslant n\}.$$

Suppose that $\sum_{n=0}^{\infty} \mathrm{var}_{n,\delta} A < +\infty$; then A satisfies condition (S) with $k = 2\sum_{n=0}^{\infty} \mathrm{var}_{n,\delta} A$. For Smale spaces this holds in particular if A is Hölder continuous [because $d(x, y) < C\lambda^n$ when $d(f^k x, f^k y) < \delta$ for $|k| < n$; see Section 7.3].

In the special case of a $\mathbf{Z}$-lattice system (see Example 7.2), A_Φ satisfies (S) when $\Phi \in \mathcal{B}_1$.

7.15 Conjugate points and conjugating homeomorphisms

Let again Ω be a general Smale space. We say that $x, y \in \Omega$ are *conjugate* if

$$\lim_{|k|\to\infty} d(f^k x, f^k y) = 0.$$

One can thus find $n > 0$ such that

$$f^n y \in V^+_{f^n x}(\delta), \qquad f^{-n} y \in V^-_{f^{-n} x}(\delta);$$

and for z in some neighborhood O of x one can define

$$\varphi z = [f^{-n}[f^n[z, x], f^n y], f^n[f^{-n} y, f^{-n}[x, z]]].$$

φ is such that $\varphi x = y$ and

(a) *φ is continuous in O,*
(b) $\lim_{|k|\to\infty} d(f^k z, f^k \varphi z) = 0$ *uniformly in z.*

We express (a), (b) by saying that (O, φ) is a *conjugating map*. Let (O', φ') be another conjugating map such that $y \in O'$, $\varphi' y = x$. Then $(O \cap \varphi^{-1} O', \varphi' \varphi)$ is a conjugating map, and it follows from (a), (b) that $\varphi' \varphi$ is the identity in some neighborhood of x. One can thus replace O by a smaller open neighborhood of x such that φ is a homeomorphism. We say then that (O, φ) is a *conjugating homeomorphism*.

The following facts are now clear.

If x, y are conjugate, there is a conjugating homeomorphism (O, φ) such that $x \in O, \varphi x = y$. Let $(O, \varphi), (O', \varphi')$ be conjugating homeomorphisms. If $O \cap \varphi^{-1} O' \neq \emptyset$, then $(O \cap \varphi^{-1} O', \varphi' \varphi)$ is a conjugating homeomorphism. If $x \in O \cap O'$ and $\varphi x = \varphi' x$, then $\varphi = \varphi'$ on a neighborhood of x.

7.16 Proposition

(a) *If x is non-wandering and y is conjugate to x, then y is non-wandering.*
(b) *If $x \in \boldsymbol{\Omega}^{\alpha\beta}$, then $\boldsymbol{\Omega}^{\alpha\beta}$ is the closure of the set of points conjugate to x.*

To prove (a) we may assume that x is periodic [because of the existence of conjugating homeomorphisms (Section 7.15) and Anosov's closing lemma (Section 7.3)]. Let x have period p, and O be a neighborhood of y. By (7.4) one can choose δ_4 so small that

$$V_u^+(\delta) \cap V_v^-(\delta) \neq \emptyset$$

if $d(u, v) < \delta_4$. We take an arbitrary large $m > 0$ such that

$$d(f^{mp}y, x) < \delta_4/2, \qquad d(f^{-mp}y, x) < \delta_4/2,$$

and, using (SS2),

$$f^{-mp}V^-_{f^{mp}y}(\delta) \subset O, \qquad f^{mp}V^+_{f^{-mp}y}(\delta) \subset O.$$

Then

$$f^{mp}O \cap f^{-mp}O \supset V^-_{f^{mp}y}(\delta) \cap V^+_{f^{-mp}(y)}(\delta) \neq \emptyset,$$

proving (a).

The sets $\boldsymbol{\Omega}^{\alpha\beta}$ have mutual distances > 0, and are invariant under f^N for suitable $N > 0$. Therefore if y is conjugate to $x \in \boldsymbol{\Omega}^{\alpha\beta}$, then $y \in \boldsymbol{\Omega}^{\alpha\beta}$ because $\lim_{|k|\to\infty} d(f^{kN}x, f^{kN}y) = 0$. To prove (b) it suffices to show that the set of points conjugate to x is dense in $\boldsymbol{\Omega}^{\alpha\beta}$. The points $x, y \in \boldsymbol{\Omega}^{\alpha\beta}$ are conjugate if and only if $\lim_{|k|\to\infty} d(f^{kn_\alpha}x, f^{kn_\alpha}y) = 0$, i.e., if and only if they are conjugate for $(\boldsymbol{\Omega}^{\alpha\beta}, f^{n_\alpha})$. All we have to prove is thus that the points conjugate to x are dense in $\boldsymbol{\Omega}$ when f is topologically mixing. It suffices to show that if $\pi\xi = x$, the points conjugate to ξ are dense in $\boldsymbol{\Omega}$: this is the property (D*) of Remark 1.14 and it holds because $(\boldsymbol{\Omega}_0, t)$ is mixing (Theorem 7.6(e)).

7.17 Theorem

Let $(\boldsymbol{\Omega}, f)$ be topologically + transitive, and ρ_0 be the f-invariant probability measure which realizes the maximum of the entropy (Bowen measure).

(a) *For each $x \in \boldsymbol{\Omega}$ there are positive measures σ^- on $V_x^-(\delta)$ and σ^+ on $V_x^+(\delta)$ such that the image of $\sigma^- \times \sigma^+$ by $[\cdot,\cdot] : V_x^-(\delta) \times V_x^+(\delta) \to \boldsymbol{\Omega}$ is the restriction of ρ_0 to $[V_x^-(\delta), V_x^+(\delta)]$.*
(b) *If (O, φ) is a conjugating homeomorphism, the image by φ of ρ_0 restricted to O is ρ_0 restricted to φO:*

$$\varphi(\rho_0|O) = \rho_0|\varphi O.$$

(a) is proved by Sinai [1], and by Ruelle and Sullivan [1], Theorem 1. It corresponds to the fact that $\rho_0 = \pi\rho_0$, where (ρ_0, τ) is a mixing Markov process when $(\mathbf{\Omega}, f)$ is topologically mixing (Parry's theorem: Parry [1]).

In proving (b) we may assume that O is a small neighborhood of x, use (a), and check that the factorizations of ρ_0 near x and φx correspond by the natural maps $V_x^{\pm} \to V_{\varphi x}^{\pm}$.

7.18 Gibbs states

Let $A \in \mathcal{C}^\alpha(\mathbf{\Omega})$. If (O, φ) is a conjugating homeomorphism define $g : O \to \mathbf{R}$ by

$$g(z) = \exp \sum_{k=-\infty}^{\infty} [A(f^k\varphi z) - A(f^k z)].$$

Since $d(f^k z, f^k\varphi z)$ tends to zero in a uniform exponential way, the function g is continuous. One can say that a probability measure $\boldsymbol{\sigma}$ on $\mathbf{\Omega}$ is a *Gibbs state for A* if

$$\varphi\,[g \cdot (\sigma|O)] = \sigma|\varphi O$$

for all conjugating homeomorphisms (O, φ). This formula expresses that the image by φ of the restriction of σ to O multiplied by g is σ restricted to φO.

If $\boldsymbol{\sigma}$ is an equilibrium state for A, then it is a Gibbs state for A. [Since $\boldsymbol{\sigma}$ is f-invariant, its support is in the non-wandering set (Section 7.4). We also know that a point conjugate to a point in a basic set is in the same basic set (Proposition 7.16(b)). Therefore it suffices to prove the statement when $(\mathbf{\Omega}, f)$ is topologically + transitive. The statement holds for $A = 0$ (Theorem 7.17(b)) and therefore for all $A \in \mathcal{C}^\alpha(\mathbf{\Omega})$ by Corollary 7.13(b).]

7.19 Periodic points

Let Fix$F$ denote the set of fixed points under a map F. If A is a continuous real function on $\mathbf{\Omega}$, and n an integer > 0, we define a partition function

$$Z_n(A) = \sum_{x \in \mathrm{Fix}\, f^n} \exp \sum_{k=0}^{n-1} A(f^k x).$$

Since f is expansive with expansive constant δ (Section 7.3), Fixf^n is *$([1, n], \delta)$-separated.*

Let $(\mathbf{\Omega}, f)$ be topologically mixing. Given $\alpha > 0$ one can find m such that if $x, y \in \mathbf{\Omega}$, there is z such that

$$d(y, z) < \alpha, \qquad d(x, f^m z) < \alpha.$$

[Let Ω be covered by finitely many balls of radius $\alpha/2$ centered at periodic points y_i. All y_i are in $\mathrm{Fix} f^N$ for some $N > 0$. Since $\cup_{n\geqslant 0} f^{nN} V^-_{y_i}(\alpha/2)$ is dense in Ω (Proposition 7.16(b)), there exists n such that every $x \in \Omega$ is at distance $< \alpha$ from $f^{nN} V^-_{x_i}(\alpha/2)$ for each i.] In view of the statements (a) and (b) of Section 7.3, one can thus choose m such that if $n > m$ and $x \in \Omega$ there is $x' \in \mathrm{Fix} f^n$ such that $d(f^k x, f^k x') < \delta$ for all $k \in [1, n-m]$. In other words $\mathrm{Fix} f^n$ *is* $([1, n-m], \delta)$*-spanning.*

Since $\mathrm{Fix} f^n$ is $([1, n], \delta])$-separated and $([1, n-m], \delta)$-spanning when f is mixing, it follows from Section 6.7 that

$$P(A) = \lim_{n\to\infty} \frac{1}{n} \log \sum_{x\in \mathrm{Fix} f^n} \exp \sum_{k=0}^{n-1} A(f^k x)$$

if $A \in \mathcal{C}$. Without the mixing assumption it follows from Smale's spectral decomposition that

$$P(A) = \limsup_{n\to\infty} \frac{1}{n} \log \sum_{x\in \mathrm{Fix} f^n} \exp \sum_{k=0}^{n-1} A(f^k x). \tag{7.11}$$

7.20 Theorem

Let (Ω, f) be topologically mixing.

(a) *The functions P_n on $\mathcal{C} = \mathcal{C}(\Omega)$ defined by*

$$P_n(A) = \frac{1}{n} \log \sum_{x\in \mathrm{Fix} f^n} \exp \sum_{k=0}^{n-1} A(f^k x)$$

tend pointwise to P when $n \to \infty$.

(b) *Define an f-invariant probability measure σ_n with support in $\mathrm{Fix} f^n$ by*

$$\sigma_n\{y\} = \frac{\exp \sum_{k=0}^{n-1} A(f^k y)}{\sum_{x\in \mathrm{Fix} f^n} \exp \sum_{k=0}^{n-1} A(f^k x)}$$

when $y \in \mathrm{Fix} f^n$. If there is a unique equilibrium state ρ_A for A, then σ_n tends to ρ_A in the vague topology when $n \to \infty$. This is in particular the case if A is Hölder continuous, or satifies conditions (S) *of Section 7.14.*

(a) was proven above. Since we have

$$\frac{d}{ds} P_n(A + sB)\bigg|_{s=0} = \sigma_0(B), \qquad \frac{d}{ds} P(A + sB) = \rho_A(B),$$

(b) follows from (a) and the convexity of the P_n.

7.21 Study of periodic points by symbolic dynamics

Periodic points for f can be studied by symbolic dynamics. In fact $\xi \in \Omega$ *is periodic if and only if* $\pi\xi$ *is periodic* (not necessarily of the same period): this is an immediate corollary of Theorem 7.6(d). It is also known that *if* $\pi\xi = \pi\eta$ is *periodic and* $\xi_0 = \eta_0$, *then* $\xi = \eta$ [Bowen [2], Proposition 12].

7.22 Proposition

There are finitely many **Z**-*lattice systems* (Ω_i, t_i) *corresponding to shifts* τ_i *acting on spaces* Ω_i, *and there are continuous maps* $\pi_i : \Omega_i \to \mathbf{\Omega}$ *and numbers* $s_i = \pm 1$ *such that the following hold.*

(a) π_i *is Lipshitz, and* $\pi_i \tau_i = f_{\pi_i}$.
(b) *There is one value, say* 1, *of the index i, such that* $s_1 = +1$, *and* τ_1, Ω_1, π_1 *are the shift, space, and map* τ, Ω, π *associated with a Markov partition.*
(c) *If* $i \neq 1$, *then* $\pi_i \Omega_i \neq \mathbf{\Omega}$.
(d) *For each* $x \in \mathbf{\Omega}$,

$$\mathrm{card}[\{x\} \cap \mathrm{Fix}\, f^n] = \sum_i s_i \mathrm{card}\left[\pi_i^{-1}\{x\} \cap \mathrm{Fix}\, \tau_i^n\right].$$

The shifts τ_i are explicitly constructed by Manning [1] so that (d) holds; (a), (b), (c) follow from the construction (see Section 7.7 for (a)).

7.23 Zeta functions

Consider the formal power series

$$\zeta(z) = \zeta_f(z) = \exp \sum_{n=1}^{\infty} |\mathrm{Fix}\, f^n| \frac{z^n}{n}.$$

Because of the above proposition

$$\zeta(z) = \prod_i \left[\exp \sum_{n=1}^{\infty} |\mathrm{Fix}\, \tau_i^n| \frac{z^n}{n}\right]^{s_i}.$$

One checks readily that

$$\left|\mathrm{Fix}\, \tau_i^n\right| = \mathrm{tr}\, t_i^n;$$

and therefore

$$\begin{aligned}\zeta(z) &= \prod_i [\exp \mathrm{tr}(-\log(1 - zt_i))]^{s_i} \\ &= \prod_i [\det(1 - zt_i)]^{-s_i},\end{aligned}$$

proving that ζ has non-vanishing convergence radius, and that *it extends to a rational function of z.*

More generally, if $A \in \mathcal{C}$, define

$$\zeta_f(ze^A) = \exp \sum_{n=1}^{\infty} \frac{z^n}{n} \sum_{x \in \operatorname{Fix} f^n} \exp \sum_{k=0}^{n-1} A(f^k x)$$
$$= \exp \sum_{n=1}^{\infty} Z_n(A) \frac{z^n}{n}.$$

If $A = 0$ this reduces to the previous ζ-function. In general the series

$$\sum_{n=1}^{\infty} Z_n(A) \frac{z^n}{n}$$

converges for $|z| < e^{-P(A)}$ because of (7.11). Therefore $z \mapsto \zeta_f(ze^A)$ is analytic and does not vanish when $|z| < e^{-P(A)}$.

Notice that $\zeta_f(ze^A)$ is the product, over the basic sets Ω^α of Smale's spectral decomposition, of the functions

$$\zeta_{f|\Omega^\alpha}(ze^{A|\Omega^\alpha}) = \zeta_{f^{n_\alpha}|\Omega^{\alpha\beta}} \left(z^{n_\alpha} \exp \sum_{m=0}^{n_\alpha - 1} A \circ f^m | \Omega^{\alpha\beta} \right),$$

so that it is possible to restrict attention to the case where (Ω, f) is topologically mixing.

By Proposition 7.22 we have then

$$\zeta_f(ze^A) = \exp \sum_{n=1}^{\infty} \frac{z^n}{n} \sum_i s_i \sum_{\xi \in \operatorname{Fix} \tau_i^n} \exp \sum_{k=0}^{n-1} A(f^k \pi_i \xi)$$
$$= \prod_i \left[\exp \sum_{n=1}^{\infty} \frac{z^n}{n} \sum_{\xi \in \operatorname{Fix} \tau_i^n} \exp \sum_{k=0}^{n-1} A \circ \pi_i (\tau_i^k \xi) \right]^{s_i}$$
$$= \prod_i [\zeta_{\tau_i}(ze^{A \circ \pi_i})]^{s_i}.$$

As noticed above, the factor

$$[\zeta_{\tau_i}(ze^{A \circ \pi_i})]^{s_i}$$

is analytic and non-zero when $|z| < e^{-P(A \circ \pi_i)}$.

By Theorem 7.9, $P(A \circ \pi_1) = P(A)$. If $i \neq 1$, we have (using Theorem 7.6(d))

$$Z_n(A \circ \pi_i) \leqslant d \cdot Z_n(A | \pi_i \Omega_i),$$

and $P(A \circ \pi_i) \leqslant P(A|\pi_i\Omega_i)$. Since $\pi_i\Omega_i$ is a closed f-invariant subset of $\mathbf{\Omega}$ and $\pi_i\Omega_i \neq \mathbf{\Omega}$ (Proposition 7.22(c)), it follows that $P(A|\pi_i\mathbf{\Omega}) < P(A)$ (Corollary 7.10(b)).

7.24 Theorem

Let $(\mathbf{\Omega}, f)$ be topologically mixing and $A \in \mathcal{C}^\alpha(\mathbf{\Omega})$. There is $R(A) > \exp[-P(A)]$ such that $z \mapsto \zeta(ze^A)$ is meromorphic, without zero, and with only one pole in the disk $\{z : |z| < R(A)\}$. The pole is simple and located at $\exp[-P(A)]$.

It remains to prove this for $\zeta_\tau(ze^{A \circ \pi})$ instead of $\zeta_f(ze^A)$, and this has been done in Theorem 5.29.

Notice that *for a general Smale space $\mathbf{\Omega}$, and $A \in \mathcal{C}^\alpha(\mathbf{\Omega})$, there is $R(A) > \exp[-P(A)]$ such that $z \mapsto \zeta(ze^A)$ is meromorphic in the disk $\{z : |z| < R(a)\}$, without zero, with poles located at $e^{-P(A)}$ times a root of unity.*

Notice also that $z \mapsto \zeta(ze^A)$ does not in general extend to a meromorphic function in $\mathbf{C}$ (see Remark 5.30).

7.25 Corollary

Let (Ω, f) be topologically mixing.

(a) *If $A \in \mathcal{C}^\alpha(\mathbf{\Omega})$, the series*

$$\sum_{n=1}^{\infty} Z_n(A)z^n - \sum_{n=1}^{\infty} e^{nP(A)}z^n$$

converges for $|z| < R(A)$; therefore

$$Z_n(A)e^{-nP(A)} - 1$$

tends to zero exponentially fast when $n \to \infty$.

(This improves part (a) of Theorem 7.20 when $A \in \mathcal{C}^\alpha(\mathbf{\Omega})$.)

(b) *The rational function extending $z \mapsto \zeta(z)$ to $\mathbf{C}$ has a simple pole at $e^{-P(0)}$, where $P(0)$ is the topological entropy, and every other pole or zero has an absolute value strictly larger than $e^{-P(0)}$.*

7.26 Expanding maps

Let Ω be a compact metric space, with metric d, and f a continuous map of Ω onto itself*. We say that f is expanding if there exist $\varepsilon > 0$ and $\lambda \in (0, 1)$ such that the following property holds:

(E) If $d(fx, y') < 2\varepsilon$, there exists a unique y such that $fy = y'$ and $d(x, y) < 2\varepsilon$. Furthermore

$$d(x, y) \leqslant \lambda d(fx, fy).$$

(E) implies that

$$d(fx, fy) \geqslant \lambda^{-1} d(x, y)$$

whenever $d(x, y) < 2\lambda\varepsilon$.

We define a map

$$\gamma : \{(x, y') \in \Omega \times \Omega : d(fx, y') < \varepsilon\} \mapsto \Omega$$

by the conditions

$$f\gamma(x, y') = y', \qquad d(x, \gamma(x, y')) < 2\varepsilon.$$

In view of (E) this defines γ uniquely and we have

$$d(x, \gamma(x, y')) \leqslant \lambda d(fx, y').$$

Furthermore γ is continuous. [Let$(x, y') \mapsto (x_0, y_0')$ and $\gamma(x, y') = y$, $\gamma(x_0, y_0') = y_0$. We have

$$d(y_0, y) \leqslant d(y_0, x_0) + d(x_0, x) + d(x, y) < 2\lambda\varepsilon$$

because $d(y_0, x_0) + d(x, y) < \lambda d(fx_0, y_0') + \lambda d(fx, y') < 2\lambda\varepsilon$ and $d(x_0, x)$ is arbitrarily small. Therefore (E) gives $d(y_0, y) \leqslant \lambda d(y_0', y') \mapsto 0$.] In particular, f is a local homeomorphism (and therefore an open map).

We let

$$\underline{\Omega} = \{(x_n)_{n\in\mathbf{Z}_\leqslant} : x_n \in \Omega \quad \text{and } fx_{n-1} = x_n \text{ for all } n\},$$
$$\mathbf{f}(x_n) = (fx_n), \qquad \mathbf{f}^{-1}(x_n) = (x_{n-1}),$$
$$\pi(x_n) = x_0.$$

This agrees with the definitions of $\underline{\Omega}$, τ, π in Section 6.17. $\underline{\Omega}$ is a compact

* If f is not onto, replace Ω by $\cap_{n\geqslant 0} f^n\Omega$.

metric space with respect to the metric

$$d((x_n),(y_n)) = \sup_{n\leqslant 0} \lambda^{|n|} d(x_n, y_n).$$

$\mathbf{f}$ is a homeomorphism of $\underline{\mathbf{\Omega}}$ such that $f\pi = \pi\mathbf{f}$, and π is a continuous and open map of $\underline{\mathbf{\Omega}}$ onto $\mathbf{\Omega}$.

If $d((x_n),(y_n)) < \varepsilon$, define $(z_n) = [(x_n),(y_n)]$ by $z_0 = x_0$ and $d(z_n, y_n) < \varepsilon$ for all n. Equivalently,

$$z_0 = x_0, \qquad z_{n-1} = \gamma(y_{n-1}, z_n).$$

The map $[\cdot,\cdot]$ is continuous because γ is, and it is readily seen that (SS1) holds; (SS2) holds also (with $\delta = \varepsilon$). Finally condition (SS3) of Section 7.11 is satisfied:

$$d((y_n),[(x_n),(y_n)]) = d(x_0, y_0) \leqslant d((x_n),(y_n));$$

and we have also

$$d(\mathbf{f}^{-1}(x_n), \mathbf{f}^{-1}(y_n)) \leqslant \lambda^{-1} d((x_n),(y_n)). \tag{7.12}$$

The space $\underline{\mathbf{\Omega}}$ (with map $[\cdot,\cdot]$ and homeomorphism $\mathbf{f}$) is thus a Smale space canonically associated with the expanding map f.

7.27 Remarks

(a) Since the construction of $\underline{\mathbf{\Omega}}$, $\mathbf{f}$, π is the same as that performed in Section 6.17, π induces a bijection $\underline{\sigma} \mapsto \sigma$ of the $\mathbf{f}$-invariant states on $\underline{\mathbf{\Omega}}$ to the f-invariant states on $\mathbf{\Omega}$ such that $h(\underline{\sigma}) = h(\sigma)$. If $A \in \mathcal{C}(\mathbf{\Omega})$ we also have $P(A\circ\pi) = P(A)$ (Section 6.18).

(b) π is a bijection of $\operatorname{Fix}\mathbf{f}^n$ to $\operatorname{Fix} f^n$ [the proof is immediate].

(c) π is a contraction. Therefore if A is Hölder continuous on $\mathbf{\Omega}$, then $A\circ\pi$ is Hölder continuous on $\underline{\mathbf{\Omega}}$.

(d) f is topologically + transitive (respectively, mixing) if and only if $\mathbf{f}$ is. [$\mathbf{f}$ is topologically + transitive if and only if, given non-empty open sets $U, V \subset \mathbf{\Omega}$, $p, q \geqslant 0$, and $N \geqslant 0$, there exists $n > N$ such that

$$\mathbf{f}^n(\mathbf{f}^p\pi^{-1}U) \cap (\mathbf{f}^q\pi^{-1}V) \neq \emptyset,$$

or if there exists $n > N + p - q$ such that

$$(\pi^{-1}U) \cap (\mathbf{f}^{-n}\pi^{-1}V) \neq \emptyset;$$

but $\mathbf{f}^{-n}\pi^{-1}V = (\pi\mathbf{f}^n)^{-1}V = (f^n\pi)^{-1}V = \pi^{-1}(f^n)^{-1}v$, so that the condition becomes

$$(\pi^{-1}U) \cap (\pi^{-1}(f^n)^{-1}V) \neq \emptyset,$$

or

$$U \cap (f^n)^{-1}V \neq \emptyset \quad \text{or} \quad f^n U \cap V \neq \emptyset.$$

The mixing case is analogous.]

7.28 Results for expanding maps

In view of the above remarks, the theory of the pressure and of equilibrium states for a Smale space extends to the situation where an expanding map is given. In particular, if f is topologically + transitive and $A \in \mathcal{C}^\alpha(\mathbf{\Omega})$, there is a unique equilibrium state ρ_A, i.e. a unique f-invariant state which makes

$$h(\rho) + \rho(A)$$

equal to its maximum $P(A)$. Similarly, the reader may check that most results mentioned in Corollaries 7.10, 7.12, 7.13 apply here directly or with simple modifications.

Using Remark 7.27(b) on periodic points, it is seen that Theorems 7.20, 7.24, and Corollary 7.25 remain true in the present case.

Further results are obtained by using a Markov partition of $\mathbf{\Omega}$.

7.29 Markov partitions

Let (R_i) be a Markov partition of $\underline{\mathbf{\Omega}}$. Each R_i is of the form $[C_i, D_i]$ where C_i is the closure of its interior in $V_{x_i}^-(\delta)$ for some x_i. Each set $\pi C_i \subset \mathbf{\Omega}$ has thus dense interior. We define sets $S_j \subset \mathbf{\Omega}$ to be the closures of the minimal non-empty intersections of the sets π int C_i. Therefore the S_j are closed and non-empty, they have dense interior, and they cover $\mathbf{\Omega}$. Furthermore

(a) $\text{int}\, S_i \cap \text{int}\, S_j = \emptyset$ if $i \neq j$;
(b) each $f S_i$ is a union of sets S_j.

We call such a family (S_i) a *Markov partition** for the expanding map f.

* Markov partitions for expanding maps can also be obtained directly by a simplification of the construction of Markov partitions for Smale spaces; cf. Bowen [1].

$(\pi^{-1}S_i)$ is formally a Markov partition of $\underline{\mathbf{\Omega}}$ (the sets $\pi^{-1}S_i$ do not have small diameter, but we may assume that for some $N > 0$, $(\mathbf{f}^N\pi^{-1}S_i)$ is a true Markov partition of $\underline{\mathbf{\Omega}}$). Therefore there is a symbolic dynamics. More precisely, let Ω_0 be the set of S_i, and write

$$t_{S_i S_j} = \begin{cases} 1 & \text{if} \quad \mathrm{int}S_i \cap f^{-1}\mathrm{int}S_j \neq \emptyset, \\ 0 & \text{otherwise.} \end{cases}$$

We define $\Omega_{\geqslant}$ as in Section 5.8, and the one-sided shift $\tilde{\tau} : \Omega_{\geqslant} \mapsto \Omega_{\geqslant}$ by $\tilde{\tau}(\xi_n)_{n\geqslant 0} = (\xi_{n+1})_{n\geqslant 0}$. This one-sided shift is related in obvious manner to the shift τ on Ω.

7.30 Theorem

If $\xi = (\xi_n)_{n\geqslant 0} \in \Omega_{\geqslant}$, then $\cap_{n\geqslant 0} f^{-n}\xi_n$ consists of a single point $\pi(\xi)$. Furthermore

(a) *The map $\pi : \Omega_{\geqslant} \mapsto \mathbf{\Omega}$ is continuous onto.*
(b) $\pi \circ \tilde{\tau} = f \circ \pi$.
(c) π^{-1} *is uniquely defined on the residual set $\mathbf{\Omega} \setminus \cup_{n\geqslant 0} f^{-n}\partial$, where $\partial = \cup_i (S_i \backslash \mathrm{int}\, S_i)$.*
(d) *There is an integer d such that, for all x, $\pi^{-1}x$ has at most d elements.*
(e) *If f is topologically + transitive (respectively mixing), then the shift τ is transitive (respectively, mixing).*

Everything follows readily from Theorem 7.6, using Remark 7.27(d) for (e).

7.31 Applications

Let f be an expanding map of $\mathbf{\Omega}$, and A a real continuous function. A map $\mathfrak{L}_A : \mathcal{C}(\mathbf{\Omega}) \mapsto \mathcal{C}(\mathbf{\Omega})$ is defined by

$$(\mathfrak{L}_A B)(x) = \sum_{y: fy = x} e^{A(y)} B(y),$$

Let $D \subset \mathbf{\Omega}$ be the residual set of points for which π^{-1} is uniquely defined. If $\pi\xi \in D$, we have

$$\begin{aligned}(\mathfrak{L}_A B)(\pi\xi) &= \sum_{\eta: \tilde{\tau}\eta = \xi} e^{A\circ\pi(\eta)} B \circ \pi(\eta) \\ &= (\mathfrak{L}(\circ\pi))(\xi),\end{aligned}$$

where $\mathfrak{L}$ is an operator on $\mathcal{C}(\Omega_{\geqslant})$ of the type defined in Section 5.11. One can apply or adapt the arguments of Chapter 5 to the present situation.

We consider an example. *If f is topologically mixing and $A \in \mathcal{C}^\alpha(\Omega)$, there exist $\tilde{A} \in \mathcal{C}^\alpha(\Omega)$ and a probability measure σ such that*

$$\operatorname{supp}\sigma = \Omega$$

and

$$\lim_{n\to\infty} e^{-nP(A)}\mathfrak{L}_A^n B = \sigma(B)\cdot e^{\tilde{A}}$$

uniformly on Ω for each $B \in \mathcal{C}(\Omega)$.

[The limit is uniform on D by Proposition 5.16, and hence on Ω by the density of D. Suppose that $|C(x') - C(x)| \leqslant cd(x,x')^\alpha$ for $d(x,x') < 2\varepsilon$; then

$$\frac{(\mathfrak{L}_A e^C)(x')}{(\mathfrak{L}_A e^C)(x)} = \frac{\sum\limits_{y':fy'=x'} \exp[A(y') + C(y')]}{\sum\limits_{y:fy=x} \exp[A(y) + C(y)]}$$
$$\leqslant \exp[a(\lambda d(x,x'))^\alpha + c(\lambda d(x,x'))^\alpha],$$

so that $|\tilde{A}(x) - \tilde{A}(x')| \leqslant \tilde{a}d(x,x')^\alpha$, where $\tilde{a} = a\lambda^\alpha(1-\lambda^\alpha)^{-1}$.]

As another example suppose that the map f has a jacobian e^A with respect to a probability measure ρ on Ω. This means that when $B \in \mathcal{C}(\Omega)$,

$$[f(B\cdot\rho)](dx) = \left[\sum_{y:fy=x} e^{A(y)}B(y)\right]\cdot\rho(dx).$$

Then

$$f^n(B\cdot\rho) = (\mathfrak{L}_A^n B)\cdot\rho.$$

In particular if f is topologically mixing and $A \in \mathcal{C}^\alpha(\Omega)$ as above, then $P(A) = 0$ and $f^n\rho \mapsto e^{\tilde{A}}\rho$, the convergence being in norm and exponentially fast [see Exercise 4(b) of Chapter 5]. In particular, $e^{\tilde{A}}\rho$ is f-invariant and equivalent to ρ.

The above argument often permits us to show that $g^n\sigma$ has a limit when σ is a measure on a compact set Λ but $g : \Lambda \mapsto \Lambda$ does not satisfy our condition (E). It suffices to find a surjective map $\omega : \Omega \mapsto \Lambda$ and f expanding such that $\omega\circ f = g\circ\omega$ and ω^{-1} is uniquely defined σ-almost everywhere.

Suppose for instance that $\Lambda = \cup_{i=1}^n \Lambda_i$ and that for each i there is a set $I(i)$ and a continuous map $g_i : \Lambda_i \mapsto \cup_{j\in I(i)}\Lambda_j$ with the following properties:

(a) $\cup_i I(i) = \{1,\ldots,n\}$.
(b) The Λ_i are closed and $\sigma(\Lambda_i \cap \Lambda_j) = 0$ if $i \neq j$.
(c) The maps g_i are bijective and

$$d(x,y) \leqslant \lambda d(fx, fy).$$

(d) There is $A_i \in \mathcal{C}^\alpha(\Lambda_i)$ such that

$$g_i(\sigma|\Lambda_i)(dx) = \exp[A_i \circ g_i^{-1}x] \cdot (\sigma| \cup_{j\in I(i)} \Lambda_j)(dx).$$

Let $g : \Lambda \mapsto \Lambda$ have restriction g_i to Λ_i (g need not be well defined on $\Lambda_i \cap \Lambda_j$ for $i \neq j$). Then $g^n\sigma$ has a limit equivalent to σ when $n \to \infty$. To see this let $\boldsymbol{\Omega}$ be the set of sequences (x_n, i_n) with $n \geqslant 0$ such that $x_n \in \Lambda_{i_n}$, $i_{n+1} \in I(i_n)$, and $x_{n+1} = g_{i_n}x_n$. Let $f(x_n, i_n) = (x_{n+1}, i_{n+1})$, $\omega(x_n, i_n) = x_0$. Then $\omega \circ f = g \circ \omega$, and if we define

$$d((x_n, i_n), (y_n, j_n)) = \sup_n \lambda^n[d(x_n, y_n) + 2\varepsilon(1 - \delta_{i_n j_n})]$$

the condition (E) is satisfied. Let $\rho = \omega^{-1}\sigma$ and $A(x_n, i_n) = A_{i_0}(x_{i_0})$. Then $A \in \mathcal{C}^\alpha(\boldsymbol{\Omega})$ and the map f has jacobian e^A with respect to ρ.

Notes

Axiom A diffeomorphisms have been introduced by Smale [1], and his paper is still the best introduction to the subject. Smale's definition extends the earlier notion of Anosov diffeomorphism (see Anosov [1]). The idea of an "abstract" study of axiom A diffeomorphisms restricted to the non-wandering set (or to a hyperbolic set) is present in Bowen [1] (cf. the "Fact 1" used in that paper). Our study is based on axioms (SS1), (SS2), and we use the term Smale space for a dynamical system with these properties. The results obtained here apply to axiom A, and in particular to Anosov diffeomorphisms.

A basic tool is the existence of Markov partitions, and of symbolic dynamics, first proved by Sinai [1], [2] for Anosov diffeomorphisms. This proof was improved and extended to axiom A diffeomorphisms by Bowen [1]. Sinai [4] realized that, using symbolic dynamics, the methods of statistical mechanics permit the study of invariant measures on a manifold with an Anosov diffeomorphism. This idea extends again to axiom A diffeomorphisms (Ruelle [5], Bowen and Ruelle [1]). The present exposition follows Sinai's ideas, and Bowen's monograph [6], with addition of some new facts.

The theory of Gibbs state presented in Sections 7.15 to 7.18 corresponds to a general definition of Gibbs states given by Capocaccia [1].

The study of periodic points in Sections 7.19 to 7.25 follows Bowen [2] and Manning [1]. Theorem 7.24 was announced in Ruelle [6].

The theory of expanding maps developed in Sections 7.26 to 7.31 is an application of the theory of Smale spaces. It is more general (and thus less rich) than the theory of expanding diffeomorphisms given by Shub [1] and Hirsch [1]. The study of the iterates of $\mathfrak{L}_A$ in Section 7.31 is an extension of the Perron–Frobenius theorem, as was Proposition 5.16. For other developments see Walters [3], [4], and in another direction Lasota and Yorke [1].

Exercises

1. Prove that a Smale space has finite Hausdorff dimension.*

   [Let δ be as in Section 7.3, and cover $\mathbf{\Omega}$ with a finite number N of sets with diameters $< \delta$. For any $n > 0$, $\mathbf{\Omega}$ is covered by at most N^{2n-1} sets S_i of diameter $< C\lambda^n$. Therefore, if $\alpha > (2\log N)/|\log\lambda|$, then

$$\lim_{n\to\infty} C\lambda^n = 0,$$

$$\lim_{n\to\infty}\sum_i (\mathrm{diam}\, S_i)^\alpha \leqslant \lim_{n\to\infty} N^{2n-1}(C\lambda^n)^\alpha = \frac{C^\alpha}{N}\lim_{n\to\infty}(N^2\lambda^\alpha)^n = 0,$$

   and this implies that $\dim\mathbf{\Omega} \leqslant (2\log N)/|\log\lambda|$; see for instance Billingsley [1], Section 14.]

2. Let $(\mathbf{\Omega}, f)$ be a topologically + transitive Smale space. The following conditions on $B \in \mathcal{C}^\alpha(\mathbf{\Omega})$ are equivalent:

   (a) $\boldsymbol{\sigma}(B) = 0$ for all $\boldsymbol{\sigma} \in \boldsymbol{I}$.

   (b) $\sum_{k=0}^{n-1} B(f^k x) = 0$ for all n and all $x \in \mathrm{Fix}\, f^n$.

   (c) $B = C \circ f - C$ with $C \in \mathcal{C}(\mathbf{\Omega})$.

   In (c), C is determined up to an additive constant, and $C \in \mathcal{C}^\alpha(\mathbf{\Omega})$ if condition (SS3) of Section 7.11 holds.

   [(c) $\Rightarrow$ (a) $\Rightarrow$ (b) are trivial. To prove (b) $\Rightarrow$ (c) one chooses y such that $\Gamma = \{f^k y : k \in \mathbf{Z}\}$ is dense in $\mathbf{\Omega}$, and constructs $C|\Gamma$ as in the proof of Theorem 5.7. Let now $u, v \in \mathbf{\Omega}$, $f^k y \to u$, $f^l y \to v$, $l - k \to +\infty$. There is z with $f^{l-k} z = z$ and $d(f^m z, f^{k+m} y) < \delta$ for $m \in [0, l-k]$ (if $d(u, v)$ is sufficiently small; see Section 7.3). Thus

$$\begin{aligned} C(v) - C(u) &= \lim[C(f^l y) - C(f^k y)] \\ &= \lim\sum_{j=0}^{l-k-1} B(f^{k+j} y) = \lim\sum_{j=0}^{l-k-1}[B(f^{k+j} y) - B(f^j z)]. \end{aligned}$$

* This result, due to Rosenberg, was communicated to me by R. Bowen.

This can be majorized term by term using

$$d(f^{k+j}y, f^j z) < \text{constant} \times \max(\lambda^j, \lambda^{(l-k)-j}).$$

One can thus take the limit $l - k \to \infty$, $z \to [u, v]$:

$$C(v) - C(u) = \sum_{j=0}^{\infty}[B(f^j u) - B(f^j[u, v])] + \sum_{j=1}^{\infty}[B(f^{-j}v) - B(f^{-j}[u, v])].$$

Hence

$$|C(v) - C(u)| \leqslant \text{constant} \times [d(u, [u, v])^\alpha + d(v, [u, v])^\alpha].$$

This implies $C \in \mathcal{C}(\mathbf{\Omega})$, and $C \in \mathcal{C}^\alpha(\mathbf{\Omega})$ if (SS3) is satisfied.]

3. Using Exercise 2, show that the function C of Corollary 7.10(c) is Hölder continuous if condition (SS3) of Section 7.11 holds.
4. Let f be an expanding map.
   (a) If f is topologically mixing, the periodic points are dense in $\mathbf{\Omega}$.
   (b) f is topologically mixing if and only if for every non-empty open set $O \subset \mathbf{\Omega}$ there is an $N > 0$ such that $f^N O = \mathbf{\Omega}$.
   (c) If the periodic points are dense in $\mathbf{\Omega}$ and $\mathbf{\Omega}$ is connected, then f is topologically mixing.

   [(a): By Remark 7.27(d), f is topologically mixing; hence $\underline{\mathbf{\Omega}}$ is non-wandering, hence the periodic points are dense in $\underline{\mathbf{\Omega}}$, and therefore in $\mathbf{\Omega}$ by Remark 7.27(b). (b): Let f be topologically mixing, O open non-empty contained in $\mathbf{\Omega}$. Using (a), let $x \in O \cap \text{Fix} f^p$. Replacing $\underline{O}$ by a sufficiently small ball around x, and using (E), we have $f^p O \supset O$, so that $f^{np} O$ increases with n. If $y \in \cup_{n>0} f^{np} O$, then (E) implies that $\{z : d(z, y) < 2\varepsilon\} \subset \cup_{n>0} f^{np} O$, and since $\cup_{n>0} f^{np} O$ is dense in $\mathbf{\Omega}$ by mixing, this set is equal to $\mathbf{\Omega}$. Since the $f^{np} O$ are open and $\mathbf{\Omega}$ is compact, there is some n such that $f^{np} O = \mathbf{\Omega}$. The converse is immediate. (c): The proof is like that of (b).]

Appendix A.1

Miscellaneous definitions and results

A.1.1 Order

Let $\leqslant$ be an order relation on a set E, and $S \subset E$. There is at most one $a \in S$ such that $a \geqslant x$ for all $x \in S$ (a is the *greatest element* of S); there is also at most one $b \in S$ such that $b \leqslant x$ for all $x \in S$ (b is the *least element* of S).

If $a \in E$ and $a \geqslant x$ for all $x \in S$, a is an *upper bound* of S, and if the set of upper bounds of S has a least element, this is called the *least upper bound* or sup of S, and denoted by $\vee S$. If $b \in E$ and $b \leqslant x$ for all $x \in S$, b is a *lower bound* of S, and if the set of lower bounds of S has a greatest element, this is called the *greatest lower bound* or inf of S, and denoted by $\wedge S$.*

An ordered set E is called a *lattice* if every finite $S \subset E$ has a least upper bound and a greatest lower bound. For this it suffices to require that every two-element subset $\{x, y\}$ has a least upper bound $x \vee y$ and a greatest lower bound $x \wedge y$.

A.1.2 Residual sets

Let E be a topological space. A subset S of E is called *residual* if it contains a countable intersection of dense open sets. If E is metrizable and complete, then every residual subset is dense in E (Baire's theorem). We say that a property (of points x of E) is *generic* if it holds for all $x \in S$, S a residual subset of E.

* Note that our use of the symbols $\vee$, $\wedge$ for covers (Section 6.3) does not quite conform to these definitions.

A.1.3 Upper semi-continuity

A function f with values in $\mathbf{R} \cup \{-\infty\}$ on the topological space E is *upper semi-continuous* if for every $x \in E$ and $a > f(x)$ there is a neighborhood $\mathfrak{N}_x$ of x such that

$$y \in \mathfrak{N}_x \Rightarrow f(y) < a.$$

It is equivalent to require that, for every real a, $\{x \in E : f(x) < a\}$ is open or $\{x \in E : f(x) \geqslant a\}$ closed.

The greatest lower bound of a family of real continuous functions on E is upper semi-continuous. If E is compact, and f is upper semi-continuous, there exists $x \in E$ such that

$$f(x) = \sup_{y \in E} f(y).$$

Thus, on a compact set, an upper semi-continuous function reaches its maximum.

A.1.4 Subadditivity

Let $F(a_1, \ldots, a_\upsilon)$ be real and defined when $a_1, \ldots, a_\upsilon$ are integers > 0. We say that F is *subadditive* if

$$F(a_1, \ldots, a_k' + a_k'', \ldots, a_\upsilon) \leqslant F(a_1, \ldots, a_k', \ldots, a_\upsilon) + F(a_1, \ldots, a_k'', \ldots, a_\upsilon)$$

for all k. We have then

$$\lim_{a_1, \ldots, a_\upsilon \to \infty} \frac{F(a_1, \ldots, a_\upsilon)}{\prod_{k=1}^{\upsilon} a_k} = \inf_{a_1, \ldots, a_\upsilon} \frac{F(a_1, \ldots, a_\upsilon)}{\prod_{k=1}^{\upsilon} a_k}.$$

The limit is a real number or $-\infty$. [This is implicit in the proof of (3.17).]

Appendix A.2
Topological dynamics

Let Ω be a non-empty Hausdorff topological space, and $f : \Omega \mapsto \Omega$ a continuous map. The pair (Ω, f) is a *topological dynamical system.* A point $x \in \Omega$ is *wandering* if it has a neighborhood U such that $U \cap \bigcup_{n=1}^{\infty} f^n U = \emptyset$. A point x is *non-wandering* if, given a neighborhood U of x, and $N \geqslant 0$, there exists $n > N$ such that $f^n U \cap U \neq \emptyset$. The non-wandering points constitute the *non-wandering set* Λ; Λ is closed and $f\Lambda \subset \Lambda$. If f is a homeomorphism, the non-wandering set for f^{-1} is the same as that for f. If Ω is compact, the non-wandering set is not empty.

We say that (Ω, f), or f, is *topologically + transitive* if it satisfies the following condition.

(+T) *If U, $V \subset \Omega$ are non-empty open sets, and $N \geqslant 0$, there exists $n > N$ such that $f^n U \cap V \neq \emptyset$.*

If f is topologically + transitive, then Ω is the non-wandering set. A homeomorphism f is topologically + transitive if and only if f^{-1} is. If Ω is a compact metrizable space, (+T) is also equivalent to the following conditions.

(+T′) *There is $x \in \Omega$ such that the set of limit points of the sequence $(f^n x)_{n>0}$ is Ω.*

(+T″) *Those $x \in \Omega$ such that the set of limit points of $(f^n x)_{n>0}$ is Ω form a residual subset of Ω.*

[Let $(V_k)_{k>0}$ be a countable basis of the topology of Ω. The set of limit points of $(f^n x)$ is Ω if $\{f^{m+n}x : n > 0\}$ is dense in Ω for all $m > 0$. The set of x for which this is true is a countable intersection of open sets:

$$\bigcap_{k>0} \bigcap_{m>0} \bigcup_{n} f^{-m-n} V_k.$$

If (+T) holds, this set has a non-empty intersection with each open $U \neq \emptyset$, and hence is dense; hence (+T) $\Rightarrow$ (+T″). Clearly (+T″) $\Rightarrow$ (+T′). If U, V are non-empty open sets and (+T′) holds, then $f^m x \in U$ and $f^{m+n} x \in V$ for some $m, n > N$; hence (+T) holds.]

Notice that if the set of limit points of $(f^n x)$ is Ω, then $\{f^n x : n > 0\}$ is dense in Ω; the converse is true if Ω contains no isolated point.

We say that (Ω, f), or f, is *topologically mixing* if it satisfies the condition

(M) *If $U, V \subset \Omega$ are non-empty open sets, there exists $N \geqslant 0$ such that*

$$f^n U \cap V \neq \emptyset \qquad \text{for all} \quad n > N.$$

If f is topologically mixing, then f is topologically + transitive. A homeomorphism f is topologically mixing if and only if f^{-1} is.

We say that a homeomorphism f is *topologically transitive* if it satisfies the condition

(T) *If U, V are non-empty open sets, there exists $n \in \mathbf{Z}$ such that $f^n U \cap V \neq \emptyset$.*

A homeomorphism f is topologically + transitive if and only if f is topologically transitive, and Ω is the nonwandering set:

$$(+\mathrm{T}) \Leftrightarrow (\mathrm{T}) \quad \text{and} \quad \Omega \textit{ is the non-wandering set.}$$

[It is trivial that (+T) $\Rightarrow$ (T) and Ω is the non-wandering set. Assume now (T) and that Ω is the non-wandering set. Let U, V be non-empty open subsets of Ω, and $N \geqslant 0$. We want to prove that

$$f^m U \cap V \neq \emptyset \qquad \text{with} \quad m > N.$$

By (T) we have $f^n U \cap V \neq \emptyset$ with $n \in \mathbf{Z}$. Since any point x of $W = f^n U \cap V$ is non-wandering, there exists $m > N$ such that $W \cap f^{m-n} W \neq \emptyset$; hence $f^m U \cap V \supset W \cap f^{m-n} W \neq \emptyset$.]

If Ω is a compact metrizable space, and f a homeomorphism, (T) is equivalent to the following conditions (see Walters [2], Chapter 5, Section 2):

(T′) *There is $x \in \Omega$ such that $\{f^n x : n \in \mathbf{Z}\}$ is dense in Ω.*
(T″) *Those $x \in \Omega$ such that $\{f^n x : n \in \mathbf{Z}\}$ is dense in Ω form a residual subset of Ω.*

Appendix A.3

Convexity

A.3.1 Generalities

Let V be a real vector space. A set $S \subset V$ is *convex* if $\alpha x + (1 - \alpha)y \in S$ whenever $x, y \in S$ and $0 \leqslant \alpha \leqslant 1$. The *convex hull* of a set $S \subset V$ is the smallest convex set containing S.

Let S be convex. A function $f : S \mapsto \mathbf{R}$ is *convex* if $\{(x, t) \in V \times \mathbf{R} : x \in S$ and $t \geqslant f(x)\}$ is convex; f is *concave* if $-f$ is convex; f is *affine* if f is convex and concave. More generally if W is a real vector space, we say that $f : S \mapsto W$ is *affine* if it satisfies $f(\alpha x + (1 - \alpha)y) = \alpha f(x) + (1 - \alpha)f(y)$ whenever $x, y \in S$ and $0 \leqslant \alpha \leqslant 1$. In particular a linear map $V \to W$ is affine.

If S is a convex open subset of $\mathbf{R}^n$, then every convex function on S is continuous. If f is a real function on the open interval $(a, b) \subset \mathbf{R}$ and $d^2 f(x)/dx^2 \geqslant 0$ for $x \in (a, b)$, then f is convex.

Convexity is central in the theory of topological vector spaces (see Köthe [1]). Here we indicate only some results used in the text.

A.3.2 Hahn–Banach theorem

Let $P : V \mapsto \mathbf{R}$ be convex, and let W be a linear subspace of V. Suppose that $w : W \mapsto \mathbf{R}$ is linear and satisfies $w \leqslant P|W$. Then there exists $v : V \mapsto \mathbf{R}$ linear and such that

$$v \leqslant P, \quad v|W = w.$$

If V is a topological vector space and P is continuous, then v is continuous. (Standard application: V is normed and P is the norm.)

A.3.3 Separation theorems

We say that two subsets S, S' of the topological vector space V are *separated* by a (closed) hyperplane if there is a continuous linear function $f : V \mapsto \mathbf{R}$ and $c \in \mathbf{R}$ such that $f(x) \leqslant c$ for $x \in S$ and $f(x) \geqslant c$ for $x \in S'$. If $f(x) < c$ for $x \in S$ and $f(x) > c$ for $x \in S'$, we say that S, S' are *strictly separated.*

Let S and S' be disjoint convex sets.

(a) If S is open, then S and S' are separated by a hyperplane.
(b) If S and S' are open, then S and S' are strictly separated by a hyperplane.
(c) If V is locally convex, S compact, and S' closed, then S and S' are strictly separated by a hyperplane.

[Note that (a) and (b) are forms of the Hahn–Banach theorem].

A.3.4 Convex compact sets

Let V be a normed space. Its dual V^* (space of continuous linear functionals on V) is a Banach space with the norm

$$\sigma \mapsto \|\sigma\| = \sup_{x \in V : \|x\| \leqslant 1} |\sigma(x)|.$$

The *weak topology* on V^* is the topology of pointwise convergence of linear functionals on V; V^* with the weak topology is locally convex, and its dual is V. The closed unit ball $\{\sigma \in V^* : \|\sigma\| \leqslant 1\}$ is compact for the weak topology of V^* (this is a special case of the *Alaoglu–Bourbaki* theorem).

Let K be a convex compact subset of the locally convex topological vector space V, and let (f_α) be a commuting family of continuous affine maps $K \mapsto K$. Then the f_α have a common fixed point (*Markov–Kakutani* theorem).

A.3.5 Extremal points

Let V be a real vector space and $S \subset V$. A point $z \in S$ is *extremal* if

$$z = \alpha x + (1 - \alpha) y \qquad \text{with} \quad x, y \in S, \quad 0 < \alpha < 1$$

imply $x = y = z$.

Let K be a convex compact subset of the locally convex topological vector space V, and let $\mathfrak{E}$ be the set of its extremal points. Then the closure of the convex hull of $\mathfrak{E}$ is K (*Krein–Milman* theorem).

Let S be a subset of the locally convex topological vector space V, such that the closure K of its convex hull is compact. Then the extremal points of K are in the closure of S (theorem of *Milman*).

A.3.6 Tangent functionals to convex functions

Let V be a topological vector space and $P : V \mapsto \mathbf{R}$ be convex continuous. A linear functional $\sigma : V \mapsto \mathbf{R}$ is called *tangent* to P at x if

$$P(x + y) \geqslant P(x) + \sigma(y) \qquad \text{for all} \quad y \in V.$$

More generally, σ is called *P-bounded* if there exists $c \in \mathbf{R}$ such that

$$\sigma - c \leqslant P.$$

By the Hahn–Banach theorem, σ is then continuous, i.e. σ is in the dual V^* of V. Furthermore at each $x \in V$ the set of tangent functionals is nonempty.

Let V be a Banach space, $P : V \mapsto \mathbf{R}$ be convex continuous, and C be a closed convex cone with apex O in V. If $\sigma_0 \in V^*$ is P-bounded, $x_0 \in V$, and $\varepsilon > 0$, there is $\sigma \in V^*$ tangent to P at x with $x \in x_0 + C$,

$$||x - x_0|| \leqslant \frac{1}{\varepsilon}[P(x_0) - \sigma_0(x_0) - s(\sigma_0)].$$

and

$$\sigma(y) \geqslant \sigma_0(y) - \varepsilon||y|| \qquad (\forall y \in C), \qquad (*)$$

where we have written $s(\sigma_0) = \inf\{P(y) - \sigma_0(y) : y \in V\}$. (Theorem of *Israel* [1]. If C is a linear subspace, the condition (*) becomes $||(\sigma - \sigma_0)|C|| \leqslant \varepsilon$; if $C = V$ one recovers a theorem by *Bishop and Phelps*.)

A.3.7 Multiplicity of tangent functionals

If V is a separable Banach space and $P : V \mapsto \mathbf{R}$ is convex continuous, the set of points $x \in V$ at which there is only one tangent functional is residual (theorem of *Mazur* [1]).

If V is a separable Banach space and $P : V \mapsto \mathbf{R}$ is convex continuous, the set

$$\{\sigma \in V^* : \sigma \text{ is tangent to } P \text{ at } x\}$$

is the closed convex hull (for the weak topology) of the set of limits

$$\sigma = \lim_{n\to\infty} \sigma_n,$$

where σ_n is the only tangent functional to P at x_n and $\lim_{n\to\infty} x_n = x$ (theorem of *Lanford and Robinson* [1]).

Appendix A.4

Measures and abstract dynamical systems

A.4.1 Measures on compact sets

Let Ω be a compact space, and $\mathcal{C}(\Omega)$ the Banach space of continuous real functions on Ω with the norm

$$A \mapsto \|A\| = \sup_{x \in \Omega} |A(x)|$$

(uniform norm). The dual $\mathcal{C}(\Omega)^*$ of $\mathcal{C}(\Omega)$ consists of the *real measures* on Ω. We call *vague topology* the weak topology on $\mathcal{C}(\Omega)^*$ (see Appendix A.3.4). $\mathcal{C}(\Omega)^*$ is also a Banach space with the norm

$$\rho \mapsto \| \rho \| = \sup_{A : \|A\| \leqslant 1} |\rho(A)|.$$

If Ω' is another compact space and $f : \Omega \mapsto \Omega'$ a continuous map, we define $f : \mathcal{C}(\Omega)^* \mapsto \mathcal{C}(\Omega')^*$ by

$$(f\rho)(A) = \rho(A \circ f) \quad \text{for} \quad \rho \in \mathcal{C}(\Omega)^*, \quad A \in \mathcal{C}(\Omega'),$$

and call $f\rho$ the *image* of the measure ρ by the continuous map f.

If $A \in \mathcal{C}(\Omega)$ and $\rho \in \mathcal{C}(\Omega)^*$, we define the product $A \cdot \rho \in \mathcal{C}(\Omega)^*$ by $(A \cdot \rho)(B) = \rho(AB)$ for all $B \in \mathcal{C}(\Omega)$.

If $\rho_1, \rho_2 \in \mathcal{C}(\Omega)^*$ and $\rho_1(A) \leqslant \rho_2(A)$ whenever $0 \leqslant A \in \mathcal{C}(\Omega)$, we write $\rho_1 \leqslant \rho_2$. This is an order relation, with respect to which $\mathcal{C}(\Omega)^*$ is a lattice (see Appendix A.1.1). A measure ρ on Ω is called a *probability measure* if it. satisfies two (and thus all) of the following conditions:

(a) $\rho \geqslant 0$,
(b) $\rho(1) = 1$,
(c) $\|\rho\| = 1$.

If $\rho \in \mathcal{C}(\Omega)^*$ and $A \in \mathcal{C}(\Omega)$, we use the notation

$$\rho(A) = \int A\, d\rho = \int A(x)\rho(dx).$$

This integration with respect to ρ can be extended from $\mathcal{C}(\Omega)$ to a large class of functions, in particular to the characteristic functions of many subsets of Ω, defining the measure of these sets (*measurable sets*). Among the measurable subsets of a metrizable compact space are the *Borel* sets: they are the elements of the σ-ring generated by the compact sets. [A non-empty class of sets is a σ-ring if it is closed under symmetric differences and countable unions.] The measurable sets are precisely the sets $X \cup N$ where $N \subset Y$, X, Y, are Borel, and Y has measure zero.

The theory just sketched is that of Radon measures on compact spaces (see for instance Bourbaki [1], [2]). Radon measures on locally compact spaces are defined analogously (an example is that of Lebesgue measure on $\mathbf{R}^n$). Except when stated otherwise, measures which occur in this monograph are Radon measures.

A.4.2 Abstract measure theory

It is possible to develop an "abstract" measure theory without assuming a topology on Ω (see for instance Halmos [1]). The fundamental object is then a *measure space* $(\Omega, \mathcal{A}, \rho)$ where $\mathcal{A}$ is a family of subsets of Ω (measurable sets) and the measure ρ is a countable additive function $\mathcal{A} \mapsto \mathbf{R}$. We assume $\rho \geqslant 0$ and $\rho(\Omega) < +\infty$. Isomorphisms of measure spaces are measure preserving maps defined and one-to-one up to sets of measure zero. One can show that the measure space defined by a positive Radon measure on a metrizable compact space is a *Lebesgue space*, i.e. isomorphic to the union of an interval of $\mathbf{R}$ with the Lebesgue measure, and a countable set (finite or infinite) such that each point has a measure or "mass" > 0 (see Rohlin [1]). In particular if a probability measure ρ on a compact metrizable space is non-atomic (i.e. $\rho\{x\} = 0$ for each point x), it defines a measure space isomorphic to the unit interval $(0, 1) \subset \mathbf{R}$ with the Lebesgue measure.

A.4.3 Abstract dynamical systems

We call a quadruple $(\Omega, \mathcal{A}, \rho, \tau)$ an *abstract dynamical system* when $(\Omega, \mathcal{A}, \rho)$ is a measure space and $\tau : \Omega \mapsto \Omega$ is an invertible map which preserves $\mathcal{A}$ and ρ. We assume furthermore that $(\Omega, \mathcal{A}, \rho)$ is isomorphic to the unit interval $(0,1) \subset \mathbf{R}$ with the Lebesgue measure.

An isomorphism of two abstract dynamical systems $(\Omega, \mathfrak{A}, \rho, \tau)$ and $(\Omega', \mathfrak{A}', \rho', \tau')$ is an isomorphism $f : (\Omega, \mathfrak{A}, \rho) \mapsto (\Omega', \mathfrak{A}', \rho')$ of measure spaces, such that $f \circ \tau = \tau' \circ f$.

The *entropy*, or *Kolmogorov–Sinai invariant*, of an abstract dynamical system is defined as in Section 6.4, replacing Borel partitions by measurable partitions. In particular this definition coincides with that of Section 6.4 for the abstract dynamical system defined by a homeomorphism of a metrizable compact space and an invariant probability measure. The entropy is a number $h \geqslant 0$ or $+\infty$; it depends only on the isomorphism class of the abstract dynamical system.

A.4.4 Bernoulli shifts

Let μ be a probability measure on the finite set Ω_0. The product measure $\rho = \mu^{\otimes \mathbf{Z}}$ on $\Omega = (\Omega_0)^{\mathbf{Z}}$ is invariant under the shift τ (see Chapter 5). We call the resulting abstract dynamical system (or any abstract dynamical system isomorphic to it) a *Bernoulli shift*. Notice that the entropy is

$$h = -\sum_{\xi \in \Omega_0} \mu\{\xi\} \log \mu\{\xi\}.$$

We assume that $h > 0$.

A.4.5 Partitions

For the definition of partitions and related definitions, see Section 6.3.

Let $(\Omega, \mathfrak{A}, \rho)$ be a measure space and $\varepsilon > 0$. Let $\mathfrak{A} = (\mathfrak{A}_i)_{i \in I}$ and $\mathfrak{B} = (\mathfrak{B}_j)_{j \in J}$ be finite measurable partitions. We say that $\mathfrak{A}$ is ε*-independent* of $\mathfrak{B}$ if there is $J' \subset J$ such that $\sum'_{j \in J'} \rho(\mathfrak{B}_j) \leqslant \varepsilon$ and

$$\sum_{i \in I} |\rho(\mathfrak{A}_i \cap \mathfrak{B}_j)/\rho(\mathfrak{B}_j) - \rho(\mathfrak{A}_i)| < \varepsilon$$

if $j \notin J'$. This relation is not symmetric, but it implies the symmetric relation

$$\sum_{i \in I} \sum_{j \in J} |\rho(\mathfrak{A}_i \cap \mathfrak{B}_j) - \rho(\mathfrak{A}_i)\rho(\mathfrak{B}_j)| < 3\varepsilon,$$

and is implied by

$$\sum_{i \in I} \sum_{j \in J} |\rho(\mathfrak{A}_i \cap \mathfrak{B}_j) - \rho(\mathfrak{A}_i)\rho(\mathfrak{B}_j)| < \varepsilon^2.$$

The partition $\mathfrak{A} = (\mathfrak{A}_i)$ is called *generating* for the abstract dynamical system $(\Omega, \mathfrak{A}, \rho, \tau)$ if the sets $\tau^k \mathfrak{A}_i (k \in \mathbf{Z})$ generate $\mathfrak{A}$ (by countable intersections and unions, and up to sets of measure zero). We say that $\mathfrak{A}$ is a *weak Bernoulli* partition if, for each $\varepsilon > 0$ there exists n such that $\vee_{x\in[n,n+k]}\tau^x\mathfrak{A}$ is ε-independent of $\vee_{x\in[-k,-1]}\tau^x\mathfrak{A}$ for all $k > 0$.

A.4.6 Isomorphism theorems

Two Bernoulli shifts with the same entropy are isomorphic (theorem of *Ornstein*).

If an abstract dynamical system (with non-atomic Lebesgue measure space) has a weak Bernoulli partition which is generating, then this system is isomorphic to a Bernoulli shift (theorem of *Friedman and Ornstein*).

For more details on the isomorphism theorems, see Ornstein [1], Shields [1], Smorodinsky [1].

Appendix A.5

Integral representations on convex compact sets

This summary follows Appendix A.5 of Ruelle [3]. Proofs of the results will be found in Bourbaki [1] and a paper by Choquet and Meyer [1], referred to in what follows as [B] and [C-M]. See also Phelps [1], Lanford [1].

A.5.1 Resultant of a measure

Let V be a locally convex topological vector space and K a convex compact subset of V. The dual $\mathcal{C}(K)^*$ of $\mathcal{C}(K)$ consists of the real measures on K; we denote by $\mathfrak{M}_+$ the convex cone of positive measures and by $\mathfrak{M}_1$ the set of positive measures of norm 1 on K ($\mathfrak{M}_1$ is the set of probability measures on K). If $\rho \in K$, we denote by $\delta_\rho \in \mathfrak{M}_1$ the unit mass (=Dirac measure) at ρ.

If $m \in \mathfrak{M}_1$, there exists $\rho \in K$ such that, for all $f \in V^*$ (=dual of V),

$$f(\rho) = \int f(\sigma) m(d\sigma).$$

ρ is called the *resultant* of m ([B], p. 216, Corollaire). If $m \in \mathfrak{M}_1$ has resultant ρ, then m can be approximated vaguely (see Appendix A.4.1) by measures $m' \in \mathfrak{M}_1$ with resultant ρ and finite support (i.e., $m' = \sum_{i=1}^n \lambda_i \delta_{\rho_i}$, $\lambda_i \geqslant 0$, $\sum_{i=1}^n \lambda_i = 1$, $\rho_i \in K$, $\sum_{i=1}^n \lambda_i \rho_i = \rho$; see [B], p. 217, Proposition 3). If $m \in \mathfrak{M}_1$ has resultant ρ and the function f is affine upper semi-continuous on K, then $m(f) = f(\rho)$ ([C-M], Lemme 10).

A compact subset S of K is called a *face* of K if $\operatorname{supp} m \subset S$ whenever the resultant of the probability measure m is in S.

A.5.2 Maximal measures

Let $S \subset \mathcal{C}(K)$ be the convex cone of convex continuous functions on K. We introduce an order relation $\prec$ on $\mathfrak{M}_+$ by

$$(m_1 \prec m_2) \Leftrightarrow (\text{for all } f \in S : m_1(f) \leqslant m_2(f)).$$

If $m_1 \prec m_2$ and f is an affine continuous function, then $m_1(f) = m_2(f)$; in particular $\|m_1\| = \|m_2\|$ and (if $m_1 \in \mathfrak{M}_1$) m_1, m_2 have the same resultant. If $m \in \mathfrak{M}_1$ and $\rho \in K$,

$$(m \text{ has resultant } \rho) \Leftrightarrow (m \succ \delta_\rho).$$

We say that $m \in \mathfrak{M}_+$ is *maximal* if it is maximal for the order $\prec$. If $m \in \mathfrak{M}_+$, there is a maximal measure $\succ m$ ([C-M], Théorème 3). In particular, if $\rho \in K$, there exists a maximal measure with resultant ρ.

A.5.3 Uniqueness problem

Let us assume that K is a basis of a convex cone C with apex O in V. This means that K is the intersection of C with a closed hyperplane $H \not\ni O$ of V which meets all generating half-lines of C. This situation can always be arranged, replacing V by $\mathbf{R} \times V$ and imbedding K in $\mathbf{R} \times V$ as $\{1\} \times K$. The cone C defines an order in V (where $\xi_1 \leqslant \xi_2$ means $\xi_2 - \xi_1 \in C$); if C is a lattice for this order (see Appendix A.1.1), we say that K is a *simplex* (Choquet). This definition does not depend on the choice of C.

The following conditions are equivalent ([C-M], Théorème 11):

(a) K is a simplex.
(b) If $\rho \in K$, there is a unique maximal measure $m_\rho \succ \delta_\rho$
(i.e., every $\rho \in K$ is the resultant of a unique maximal measure m_ρ).

If K is a simplex, the mapping $\rho \mapsto m_\rho$ is affine ([C-M], proof of Théorème 11).

A.5.4 Maximal measures and extremal points

Let $\mathcal{E}(K)$ be the set of extremal points of K. If $m \in \mathfrak{M}_+$ is carried by $\mathcal{E}(K)$ (i.e. if $\mathcal{E}(K)$ is m-measurable and $m(K \backslash \mathcal{E}(K)) = 0$), then m is maximal ([C-M], Proposition 15). Conversely, if K is metrizable and $m \in \mathfrak{M}_+$ is maximal, then m is carried by $\mathcal{E}(K)$ ([C-M], Lemme 13).

Therefore if K is metrizable and $m \in \mathfrak{M}_+$,

$$(m \text{ maximal}) \Leftrightarrow (m \text{ carried by } \mathcal{E}(K)).$$

In particular every $\rho \in K$ is the resultant of a measure m_ρ carried by $\mathcal{E}(K)$, and if K is a simplex, $\rho \mapsto m_\rho$ is one-to-one from K onto the probability measures on K carried by $\mathcal{E}(K)$. In this case we may say that every $\rho \in K$ has a unique integral representation on $\mathcal{E}(K)$ by a measure m_ρ such that $f(\rho) = m_\rho(f)$ for any affine continuous function f on K.

A.5.5 Simplexes of measures

Let Ω be a compact space. We put on the space $\mathcal{C}(\Omega)^*$ of real measures on Ω the vague topology (see Appendix A.4.1). The set $E = \mathfrak{M}_1(\Omega)$ of probability measures on Ω is then compact; E is metrizable if Ω is.

Let $\mathcal{S}$ be a closed linear subspace of $\mathcal{C}(\Omega)^*$. If $\sigma \in \mathcal{S}$ implies $|\sigma| \in \mathcal{S}$, then $K = E \cap \mathcal{S}$ is a simplex. If $\rho, \rho' \in E \cap \mathcal{S}$, then $\|\rho' - \rho\| = \|m_{\rho'} - m_\rho\|$. In particular if ρ, ρ' are distinct extremal points of $E \cap \mathcal{S}$, then $\|\rho' - \rho\| = 2$, i.e. ρ and ρ' are disjoint measures. [Let $H = \{\sigma \in \mathcal{S} : \sigma(1) = 1\}$ and $\mathcal{S}_+$ be the convex cone of positive measures in $\mathcal{S}$. Since $E \cap \mathcal{S} = H \cap \mathcal{S}_+$, this set is a basis of $\mathcal{S}_+$. Let $\sigma_1, \sigma_2 \in \mathcal{S}_+$; then $\sigma_\pm = \frac{1}{2}(\sigma_1 + \sigma_2 \pm |\sigma_1 - \sigma_2|)$, which are the sup and inf of σ_1 and σ_2 in $\mathcal{C}(\Omega)^*$, are in $\mathcal{S}$ by assumption, and hence in $\mathcal{S}_+$. Therefore $\sigma_\pm$ are the sup and inf of σ_1, σ_2 for the order defined by $\mathcal{S}_+ : \mathcal{S}_+$ is a simplicial cone and $E \cap \mathcal{S}$ is a simplex. Let now $\rho, \rho' \in E \cap \mathcal{S}$. Define $\rho_\pm = \frac{1}{2}[|\rho' - \rho| \pm (\rho' - \rho)] \geqslant 0$ and let $m_\pm = \|\rho_\pm\| m_{\rho_\pm/\|\rho_\pm\|}$ or 0 if $\|\rho_\pm\| = 0$. Then m_+, m_- are disjoint (because ρ_+, ρ_- are disjoint), and $m_{\rho'} - m_\rho = m_+ - m_-$ (because $\rho' - \rho = \rho_+ - \rho_-$ and $\sigma \mapsto m_\sigma$ is affine). Thus

$$\begin{aligned}\|\rho' - \rho\| &= \|\rho_+ - \rho_-\| = \|\rho_+\| + \|\rho_-\| = \|m_+\| + \|m_-\| \\ &= \|m_+ - m_-\| = \|m_{\rho'} - m_\rho\|.]\end{aligned}$$

A.5.6 $\mathbf{Z}^\nu$-invariant measures

Let Ω be a compact space, τ a $\mathbf{Z}^\nu$-action by homeomorphisms on Ω, and $I \subset \mathcal{C}(\Omega)^*$ the simplex of τ-invariant probability measures on Ω. The unique maximal measure m_ρ on I with resultant $\rho \in I$ is determined by

$$m_\rho\left(\prod_{i=1}^{l} \hat{A}_i\right) = \lim_{\Lambda_1,\ldots,\Lambda_l \nearrow \infty} \rho\left(\prod_{i=1}^{l}\left(|\Lambda_i|^{-1} \sum_{x \in \Lambda_i} A_i \circ \tau^x\right)\right),$$

where $\hat{A} : I \mapsto \mathbf{R}$ is defined by $\hat{A}(\sigma) = \sigma(A)$.

We say that $\rho \in I$ is *ergodic* if it is an extremal point of I; this is the case if and only if $m_\rho(\hat{A}^2) = \rho(A)^2$ for all $A \in \mathcal{C}$. The integral representation $\rho \mapsto m_\rho$ is called the *ergodic decomposition*. (See Ruelle [3], Chapter 6.)

Appendix B
Open problems

In this appendix some open problems – of rather different levels of difficulty and interest – have been collected.

B.1 Systems of conditional probabilities (Chapter 2)

How general is a system of conditional probabilities $(\mu_{(\Lambda)\xi_{L\setminus\Lambda}})$ of the form $(\mu^{\Phi}_{(\Lambda)\xi_{L\setminus\Lambda}})$, where Φ is an interaction? (On this problem see in particular Sullivan [1].)

B.2 Theory of phase transitions (Chapter 3)

Show that in a suitable space of interactions the set of points of coexistence of $n + 1$ phases is a manifold of codimension n. What are the incidence relations of the manifolds thus obtained? How do critical points arise? (For a "heuristic theory" see Ruelle [8].)

B.3 Abstract measure-theory viewpoint (Chapter 4)

In Remark 4.5 a Gibbs state is obtained from another one by multiplication by a continuous function and taking a certain vague limit. Is there an abstract measure-theory version of this? In particular, can one make use of the Bernoulli property of Theorem 5.10?

B.4 A theorem of Dobrushin (Chapter 5)

Can the theorem of Dobrushin [4] on the analyticity of the pressure for one-dimensional systems be extended to mixing systems?

B.5 Definition of the pressure (Chapter 6)

If τ is expansive, the pressure can be defined by a limit for $a \to \infty$ (see end of Section 6.7). Could one use a limit for $\Lambda \nearrow \infty$? Notice that this was possible in the situation of Chapter 3 (Corollary 3.13).

B.6 Shub's entropy conjecture (Chapter 6)

Let f be a diffeomorphism of a compact manifold and f_* the corresponding linear operator on homology (with real coefficients). Is it true that the logarithm of the spectral norm of f_* is $\leqslant$ the topological entropy of f? (On this well-known conjecture see in particular Manning [2].)

B.7 The condition (SS3) (Chapter 7)

If (SS1) and (SS2) hold, can one find a metric such that (SS3) holds, and also

$$d(fx, fy) \leqslant Ld(x, y), \quad d(f^{-1}x, f^{-1}y) \leqslant Ld(x, y)$$

for some constant $L > 0$ (see Equation (12))?

B.8 Gibbs states on Smale spaces (Chapter 7)

Are Gibbs states on a Smale space (see Section 7.18) necessarily equilibrium states?

B.9 Cohomological interpretation (Chapter 7)

Can one give a cohomological interpretation of Manning's formula (Proposition 7.22) and of the rational zeta function $\zeta(z)$? (On this problem see in particular Franks [1].)

B.10 Smale flows (Chapter 7 and Appendix C)

Can one give a version of the theory of Smale spaces for flows? (On this problem see in particular Bowen [4].) See also the questions in Appendix C.4.

Appendix C

Flows

A flow on a set $\boldsymbol{\Omega}$ is a family $(\boldsymbol{\tau}^t)_{t\in\mathbf{R}}$ of maps $\boldsymbol{\tau}^t : \boldsymbol{\Omega} \mapsto \boldsymbol{\Omega}$ such that $\boldsymbol{\tau}^{t+t'} = \boldsymbol{\tau}^t \circ \boldsymbol{\tau}^{t'}$ and $\boldsymbol{\tau}^0$ is the identity. The replacement of $\mathbf{Z}$ by $\mathbf{R}$ in the thermodynamic formalism can be done in several inequivalent ways. Here we shall not consider the usual statistical mechanics of continuous one-dimensional classical systems (see Ruelle [3]), but describe the formalism appropriate to the study of flows on differentiable manifolds.

C.1 Thermodynamic formalism on a metrizable compact set

Let $\boldsymbol{\Omega}$ be compact metrizable and $(\boldsymbol{\tau}^t)$ a continuous flow, i.e., $(x, t) \mapsto \boldsymbol{\tau}^t x$ is continuous. The set $\boldsymbol{I}$ of $\boldsymbol{\tau}$-invariant probability measures on $\boldsymbol{\Omega}$ is convex and compact for the vague topology. If $\boldsymbol{\sigma} \in \boldsymbol{I}$, then

$$h_{\boldsymbol{\tau}^t}(\boldsymbol{\sigma}) = |t| h_{\boldsymbol{\tau}}(\boldsymbol{\sigma}),$$

where $h_{\boldsymbol{\tau}}(\boldsymbol{\sigma})$ is called the *(mean) entropy* of $\boldsymbol{\sigma}$ with respect to the flow $(\boldsymbol{\tau}^t)$ (see Abramov [1]).

Let d be a metric on $\boldsymbol{\Omega}$ compatible with the topology. Given $\varepsilon > 0$, $T > 0$, we say that a subset S of $\boldsymbol{\Omega}$ is *(T, ε)-separated* if $x, y \in S$ and $x \neq y$ imply

$$d(\boldsymbol{\tau}^t x, \boldsymbol{\tau}^t y) > \varepsilon \qquad \text{for some} \quad t \in [0, T].$$

Then, if $A \in \mathcal{C}(\boldsymbol{\Omega})$, we define

$$Z_T(A, \varepsilon) = \sup \left\{ \sum_{x\in S} \exp \int_0^T A(\boldsymbol{\tau}^t x) dt : S \text{ is } (T, \varepsilon)\text{-separated} \right\},$$

$$P(A) = P_{\boldsymbol{\tau}}(A) = \lim_{\varepsilon\to 0} \limsup_{T\to\infty} \frac{1}{T} \log Z_T(A, \varepsilon).$$

The definition of $P(A)$ is independent of the choice of the metric on $\boldsymbol{\Omega}$. If $A^1(x) = \int_0^1 A(\boldsymbol{\tau}^t x)dt$, one has $P(A) = P_{\boldsymbol{\tau}^1}(A^1)$. The *pressure* P satisfies the *variational principle*

$$P_{\boldsymbol{\tau}}(A) = \sup_{\boldsymbol{\sigma}\in\mathbf{I}} [h_{\boldsymbol{\tau}}(\boldsymbol{\sigma}) + \boldsymbol{\sigma}(A)]$$

(see Bowen and Ruelle [1]). The measures $\boldsymbol{\sigma}$ making $h_{\boldsymbol{\tau}}(\boldsymbol{\sigma}) + \boldsymbol{\sigma}(A)$ maximum are called *equilibrium states* for A.

C.2 Special flows

Let Ω be a compact metrizable space, $\tau : \Omega \mapsto \Omega$ a homeomorphism, and $\psi : \Omega \mapsto \mathbf{R}$ a strictly positive continuous function. In the set

$$U = \{(\xi, u) \in \Omega \times \mathbf{R} : 0 \leqslant u \leqslant \psi(\xi)\}$$

we identify $(\xi, \psi(\xi))$ and $(\tau\xi, 0)$, obtaining a compact metrizable space $\boldsymbol{\Omega}$. A continuous flow $(\boldsymbol{\tau}^t)$ on $\boldsymbol{\Omega}$ exists such that

$$\boldsymbol{\tau}^t(\xi, u) = (\xi, u + t) \qquad \text{if} \quad 0 \leqslant u + t \leqslant \psi(\xi).$$

Let σ belong to the set I of τ-invariant probability measures on Ω. If m denotes the Lebesgue on $\mathbf{R}$, then $\sigma \times m/(\sigma \times m)(U)$ defines a measure $\boldsymbol{\sigma} \in \boldsymbol{I}$ on $\boldsymbol{\Omega}$. The map $\sigma \to \boldsymbol{\sigma}$ is a bijection $I \to \boldsymbol{I}$, and

$$h_{\boldsymbol{\tau}}(\boldsymbol{\sigma}) = \frac{h_{\tau}(\sigma)}{\sigma(\psi)}$$

by a theorem of Abramov [1].

C.3 Special flow over a Smale space

With the above notation, let Ω be a Smale space, τ topologically mixing, and $\psi \in \mathcal{C}^\alpha(\Omega)$. Given $A \in \mathcal{C}(\boldsymbol{\Omega})$, let $\varphi(\xi) = \int_0^{\psi(\xi)} A(\xi, u)du$ and assume that $\varphi \in \mathcal{C}^\alpha(\Omega)$. *There is a unique equilibrium state* $\boldsymbol{\sigma}$ *on* $\boldsymbol{\Omega}$ *for* A; $\boldsymbol{\sigma}$ *corresponds to* $\sigma \in I$, *where* σ *is the unique equilibrium state for* φ-$P(A) \cdot \psi$ *on* Ω (Bowen and Ruelle [1]). *If* $(\boldsymbol{\tau}^t)$ *is topologically mixing, then* $(\boldsymbol{\sigma}, (\boldsymbol{\tau}^t))$ *is a Bernoulli flow* (Bunimovič [1], Ratner [2]).

Define

$$\zeta_A(s) = \prod_\gamma \left[1 - \exp \int_0^{\lambda(\gamma)} \left(A\left(\boldsymbol{\tau}^t x_\gamma\right) - s\right) dt\right]^{-1},$$

where the product extends over the periodic orbits γ of the flow, $\lambda(\gamma)$ is the prime period of γ, and x_γ is a point of γ. This zeta function can be rewritten

$$\zeta_A(s) = \exp \sum_{m=1}^{\infty} \frac{1}{m} \sum_{\xi \in \mathrm{Fix}\,\tau^m} \exp \sum_{k=0}^{m-1} [\varphi(\tau^k \xi) - s\psi(\tau^k \xi)].$$

ζ_A is an analytic function of s for $\mathrm{Re}\, s > P(A)$*, with a simple pole at* $P(A)$ (Ruelle [6]).

C.4 Problems

Assume that (τ^t) is topologically mixing.

(a) Let $B_1, B_2 \in \mathcal{C}^\alpha(U)$ and $\mathrm{supp}\, B_1, B_2 \subset \{(\xi, u) : 0 < u < \psi(\xi)\}$. Does $\sigma(B_1 \cdot (B_2 \circ \tau^t)) - \sigma(B_1)\sigma(B_2)$ tend to 0 exponentially fast when $|t| \to \infty$?

(b) Is there $r > 0$ such that ζ_A is meromorphic for $\mathrm{Re}\, s > P(A) - r$ with only one pole at $P(A)$?

Appendix D

Update of open problems

Since the present monograph was first published, certain topics discussed in it have received a lot of attention. A good source of information on this recent work is provided by:

W. Parry and M. Pollicott. *Zeta functions and the Periodic Orbit Structure of Hyperbolic Dynamics*. Astérisque 187–188, Soc. Math. de France, Paris, 1990, pp. 1–268.

Here we only indicate some references relevant to the list of open problems of Appendix B.

B.2 There are partially negative results:

H. A. M. Daniëls and A. C. D. van Enter. "Differentiability properties of the pressure in lattice systems," *Commun. Math. Phys.* **71**, 65–76 (1980).

A. C. D. van Enter. "Stability properties of phase diagrams in lattice systems." Thesis, Groningen, 1981.

B.6 Shub's conjecture has been proved for C^∞ diffeomorphisms:

Y. Yomdin. "Volume growth and entropy," *Israel J. Math.* **57**, 301–317 (1987).

B.7 This problem has a positive answer; Fried has shown that there is a metric d such that (SS 3) holds and f, f^{-1} are Lipschitz. In particular the function C of Corollary 7.10 (c) is Hölder continuous.

D. Fried. "Métriques naturelles sur les espaces de Smale." *C. R. Acad. Sc. Paris* **297**, Sér. I, 77–79 (1983).

Incidentally, let us mention the following reference on the theory of expanding maps:

E. M. Coven and W. L. Reddy. "Positively expansive maps of compact manifolds," in *Global Theory of Dynamical Systems*. Lecture Notes in Math. no. 819, pp. 96–110. Springer, Berlin, 1980.

B.8 This problem has a positive answer:

N. T. A. Haydn. "On Gibbs and equilibrium states," *Ergod. Th. and Dynam. Syst.* **7**, 119–132 (1987).

B10 For this problem see:

M. Pollicott. "Symbolic dynamics for Smale flows," *Amer. J. Math.* **109**, 183–200 (1987).

Problems C.4 (a) and (b) have a negative answer:

D. Ruelle. "Flots qui ne mélangent pas exponentiellement," *C. R. Acad. Sc. Paris* **296**, Sér. I, 191–193 (1983).

It has, however, been shown by Parry and Pollicott that ζ_0 is holomorphic in a neighborhood of the line Re s = P(0), except for the pole at P(0). This implies an analogue of the prime number theorem for periodic orbits of an Axiom A flow:

W. Parry and M. Pollicott. "An analogue of the prime number theorem for the closed orbits of Axiom A flows," *Ann. Math.* **118**, 573–591 (1983).

References

L. M. Abramov, "On the entropy of a flow," *Dokl. Akad. Nauk SSSR* **128**, 873–875 (1959). English translation, *Amer. Math. Soc. Transl.*, Ser. 2, **49**, 167–170 (1966).

R. L. Adler, A. G. Konheim, and M. H. McAndrew, "Topological entropy," *Trans. Amer. Math. Soc.* **114**, 309–319 (1965).

D. V. Anosov, "Geodesic flows on a compact Riemann manifold of negative curvature," *Trudy Mat. Inst. Steklov* **90** (1967). English translation, *Proc. Steklov Math. Inst.* **90** (1967).

H. Araki, "Gibbs states of a one-dimensional quantum lattice," *Commun. Math. Phys.* **14**, 120–157 (1969).

R. Berger, "The undecidability of the domino problem," *Mem. Amer. Math. Soc.*, No. 66, 1966.

P. Billingsley, *Ergodic Theory and Information.* John Wiley, New York, 1965.

N. Bourbaki, *Eléments de mathématique. Intégration. Chapitres 1, 2, 3, et 4*, 2ᵉ éd. Hermann, Paris, 1965.

Eléments de mathématique. Intégration. Chapitre 5, 2ᵉ éd. Hermann, Paris, 1967.

R. Bowen, "Markov partitions for axiom A diffeomorphisms," *Amer. J. Math.* **92**, 725–747 (1970).

"Markov partitions and minimal sets for axiom A diffeomorphisms," *Amer. J. Math.* **92**, 907–918 (1970).

"Entropy for group endomorphisms and homogeneous spaces," *Trans. Amer. Math. Soc.* **153**, 401–414 (1971).

"Symbolic dynamics for hyperbolic flows," *Amer. J. Math.* **95**, 429–459 (1973).

"Some systems with unique equilibrium states," *Math. Systems Theory* **8**, 193–202 (1974).

Equilibrium States and the Ergodic Theory of Anosov diffeomorphisms. Lecture Notes in Math. No. 470. Springer, Berlin, 1975.

R. Bowen and D. Ruelle, "The ergodic theory of axiom A flows," *Inventiones Math.* **29**, 181–202 (1975).

L. A. Bunimovič, "Imbedding of Bernoulli shifts in certain special flows," *Uspehi Mat. Nauk* **28**, No. 3, 171–172 (1973).

D. Capocaccia, "A definition of Gibbs state for a compact set with Z^ν action," *Commun. Math. Phys.* **48**, 85–88 (1976).

G. Choquet and P.-A. Meyer, "Existence et unicité des représentations intégrales dans les convexes compacts quelconques," *Ann. Inst. Fourier* **13**, 139–154 (1963).

K. L. Chung, *Markov Chains with Stationary Transition Probabilities*. Springer, Berlin, 1967.

J. P. Conze, "Entropie d'un groupe abélien de transformations," *Zeitschr. Wahrscheinlichkeitstheorie Verw. Gebiete* **25**, 11–30 (1972).

M. Denker, "Remarques sur la pression pour les transformations continues," *C. R. Acad. Sci. Paris* **279**, A967–A970 (1974).

M. Denker, C. Grillenberger, and K. Sigmund, *Ergodic Theory on Compact Spaces*. Lecture Notes in Mathematics No. 527. Springer, Berlin, 1976.

E. I. Dinaburg, "The relation between topological entropy and metric entropy," *Dokl. Akad. Nauk SSSR* **190**, No. 1, 19–22 (1970). English translation, *Soviet Math. Dok.* **11**, 13–16 (1970).

R. L. Dobrushin, "The description of a random field by means of conditional probabilities and conditions of its regularity," *Teorija Verojatn. i ee Prim.* **13**, 201–229 (1968). English translation, *Theory Prob. Applications* **13**, 197–224 (1968).

"Gibbsian random fields for lattice systems with pairwise interactions," *Funkts. Analiz i ego Pril.* **2**, No. 4, 31–43 (1968). English translation, *Functional Anal. Appl.* **2**, 292–301 (1968).

"The problem of uniqueness of a Gibbsian random field and the problem of phase transitions," *Funkts. Analiz i ego Pril.* **2**, No. 4, 44–57 (1968). English translation, *Functional Anal. Appl.* **2**, 302–312 (1968).

"Analyticity of correlation functions in one-dimensional classical systems with slowly decreasing potentials," *Commun. Math. Phys.* **32**, 269–289 (1973).

F. J. Dyson, "Existence of a phase-transition in a one-dimensional Ising ferromagnet," *Commun. Math. Phys.* **12**, 91–107 (1969).

S. A. Elsanousi, "A variational principle for the pressure of a continuous Z^2-action on a compact metric space," *Amer. J. Math.* **99**, 77–106 (1977).

M. E. Fisher, "The theory of condensation and the critical point," *Physics* **3**, 255–283 (1967).

J. M. Franks, "A reduced zeta function for diffeomorphisms," *Amer. J. Math.* **100**, No. 2 (1978).

H. Furstenberg and H. Kesten, "Products of random matrices," *Ann. Math. Statist.* **31**, 457–469 (1960).

G. Gallavotti, "Ising model and Bernoulli schemes in one dimension," *Commun. Math. Phys.* **32**, 183–190 (1973).

"Funzioni zeta ed insiemi basilari," *Accad. Lincei. Rend. Sc. fis. mat. e nat.* **61**, 309–317 (1976).

G. Gallavotti and S. Miracle-Sole, "Statistical mechanics of lattice systems," *Commun. Math. Phys.* **5**, 317–323 (1967).

F. R. Gantmaher, *The Theory of Matrices*. Nauka, Moscow, 1967. English translation, Chelsea, New York, 1964.

H.-O. Georgii, *Phasenübergang 1. Art bei Gittergasmodellen*. Lecture Notes in Physics No. 16. Springer, Berlin, 1972.

"Two remarks on extremal equilibrium states," *Commun. Math. Phys.* **32**, 107–118 (1973).

T. N. T. Goodman, "Relating topological entropy and measure entropy," *Bull. London Math. Soc.* **3**, 176–180 (1971).

L. W. Goodwyn, "Topological entropy bounds measure-theoretic entropy," *Proc. Amer. Math. Soc.* **23**, 679–688 (1969).

R. B. Griffiths and D. Ruelle, "Strict convexity ('continuity') of the pressure in lattice systems," *Commun. Math. Phys.* **23**, 169–175 (1971).

B. M. Gurevič, "Topological entropy of enumerable Markov chains," *Dokl. Akad. Nauk SSSR* **187**, No. 4, 754–757 (1969). English translation, *Soviet Math. Dokl.* **10**, 911–915 (1969).

B. M. Gurevič and V. I. Oseledec, "Gibbs distributions and dissipativeness of U-diffeomorphisms," *Dokl. Akad. Nauk SSSR* **209**, No. 5, 1021–1023 (1973). English translation, *Soviet Math. Dokl.* **14**, 570–573 (1973).

H. Halmos, *Measure Theory*. D. Van Nostrand, Princeton, 1950.

M. W. Hirsch, "Expanding maps and transformation groups," in *Global Analysis* Proc. Symp. Pure Math. **14**, 1970, pp. 125–131.

R. B. Israel, "Existence of phase transitions for long-range interactions," *Commun. Math. Phys.* **43**, 59–68 (1975).

Tangents to the Pressure as Invariant Equilibrium States in Statistical Mechanics of Lattice Systems, Princeton University Press, Princeton, 1978.

M. Keane, "Sur les mesures invariantes d'un recouvrement régulier," *C. R. Acad. Sci. Paris* **272**, A585–A587 (1971).

G. Köthe, *Topologische lineare Räume* I. Springer, Berlin, 1960.

O. E. Lanford, "Selected topics in functional analysis," in *Mécanique statistique et théorie quantique des champs. Les Houches 1970.* (C. De Witt, and R. Stora, eds.), pp. 109–214. Gordon and Breach, New York, 1971.

"Entropy and equilibrium states in classical statistical mechanics," in *Statistical mechanics and mathematical problems*, Lecture Notes in Physics No. 20, pp. 1–113. Springer, Berlin, 1973.

O. E. Lanford and D. W. Robinson, "Statistical mechanics of quantum spin systems III," *Commun. Math. Phys.* **9**, 327–338 (1968).

O. E. Lanford and D. Ruelle, "Observables at infinity and states with short range correlations in statistical mechanics," *Commun. Math. Phys.* **13**, 194–215 (1969).

A. Lasota and J. A. Yorke, "On the existence of invariant measures for piecewise monotonic transformations," *Trans. Amer. Math. Soc.* **186**, 481–488 (1973).

F. Ledrappier, "Mesures d'équilibre sur un réseau," *Commun. Math. Phys.* **33**, 119–128 (1973).

"Principe variationnel et systèmes dynamiques symboliques," *Z. Wahrscheinlichkeitstheorie Verw. Gebiete* **30**, 185–202 (1974).

F. Ledrappier and P. Walters, "A relativised variational principle for continuous transformations," *J. London Math. Soc.* **16**, 568–576 (1977).

A. N. Livšic, "Homology properties of Y-systems," *Mat. Zametki* **10**, No. 5, 555–564 (1971). English translation, *Math. Notes* **10**, 758–763 (1971).

"Cohomology of dynamical systems," *Izv. Akad. Nauk SSSR. Ser. Mat.* **36**, No. 6, 1296–1320 (1972). English translation, *Math. USSR Izvestija* **6**, 1276–1301 (1972).

A. Manning, "Axiom A diffeomorphisms have rational zeta functions," *Bull. London Math. Soc.* **3**, 215–220 (1971).

"Topological entropy and the first homology group," in *Dynamical Systems. Warwick 1974*, Lecture Notes in Mathematics No. 468, pp. 185–190. Spinger, Berlin, 1975.

S. Mazur, "Über konvexe Mengen in linearen normierten Räumen," *Studia Math.* **4**, 70–84 (1933).

M. Misiurewicz, "A short proof of the variational Principle for a Z_+^N action on a compact space," *Astérisque* **40**, 147–157 (1976).

D. S. Ornstein, *Ergodic Theory, Randomness, and Dynamical Systems.* Yale Mathematical Monographs **5**. Yale University Press, New Haven, 1974.

V. I. Oseledec, "A multiplicative ergodic theorem. Ljapunov characteristic numbers for dynamical systems," *Trudy Moscov. Mat. Obšč.* **19**, 179–210 (1968). English translation, *Trans. Moscow Math. Soc.* **19**, 197–231 (1968).

W. Parry, "Intrinsic Markov chains," *Trans. Amer. Math. Soc.* **112**, 55–66 (1964).

"Topological Markov chains and suspensions," Warwick preprint, 1974.

R. Phelps, *Lectures on Choquet's Theorem.* Van Nostrand Mathematical Studies No. 7. D. Van Nostrand, Princeton, 1966.

C. J. Preston, *Gibbs States on Countable Sets.* Cambridge Tracts in Mathematics No. 68. Cambridge University Press, Cambridge, 1974.

Random Fields. Lecture Notes in Mathematics No. 534. Springer, Berlin, 1976.

M. Ratner, "The central limit theorem for geodesic flows on n-dimensional manifolds of negative curvature," *Israel J. Math.* **16**, 181–197 (1973).

"Anosov flows with Gibbs measures are also Bernoullian," *Israel J. Math.* **17**, 380–391 (1974).

R. M. Robinson, "Undecidability and nonperiodicity for tilings of the plane," *Inventiones Math.* **12**, 177–209 (1971).

D. W. Robinson and D. Ruelle, "Mean entropy of states in classical statistical mechanics." *Commun. Math. Phys.* **5**, 288–300 (1967).

V. A. Rohlin, "On the fundamental ideas of measure theory," *Mat. Sbornik (N. S.)* **25**, 107–150 (1949). English translation, *Amer. Math. Soc. Transl.*, Ser. 1, **10**, 1–54 (1952).

D. Ruelle, "A variational formulation of equilibrium statistical mechanics and the Gibbs phase rule," *Commun. Math. Phys.* **5**, 324–329 (1967).

"Statistical mechanics of a one-dimensional lattice gas," *Commun. Math. Phys.* **9**, 267–278 (1968).

Statistical Mechanics. Rigorous Results. Benjamin, New York, 1969.

"Statistical mechanics on a compact set with Z^ν-action satisfying expansiveness and specification," *Bull. Amer. Math. Soc.* **78**, 988–991 (1972); *Trans. Amer. Math. Soc.* **185**, 237–251 (1973).

"A measure associated with axiom A attractors," *Amer. J. Math.* **98**, 619–654 (1976).

"Generalized zeta-functions for axiom A basic sets," *Bull. Amer. Math. Soc.* **82**, 153–156 (1976).

"Zeta-functions for expanding maps and Anosov flows," *Inventiones Math.* **34**, 231–242 (1976).

"A heuristic theory of phase transitions," *Commun. Math. Phys.*, **53**, 195–208 (1977).

D. Ruelle and D. Sullivan, "Currents, flows and diffeomorphisms," Topology **14**, 319–327 (1975).

P. Shields, *The Theory of Bernoulli Shifts.* University of Chicago Press, Chicago, 1973.

M. Shub, "Endomorphisms of compact differentiable manifolds," *Amer. J. Math.* **91**, 175–199 (1969).

B. Simon, *The $P(\varphi)_2$ Euclidean (Quantum) Field Theory*. Princeton University Press, Princeton, 1974.

Ia. G. Sinai, "Markov partitions and C-diffeomorphisms," *Funkts. Analiz i Ego Pril.* **2**, No. 1, 64–89 (1968). English translation, *Functional Anal. Appl.* **2**, 61–82 (1968).

"Construction of Markov partition," *Funkts. Analiz i Ego Pril.* **2**, No. 3, 70–80 (1968). English translation, *Functional Anal. Appl.* **2**, 245–253 (1968).

"Mesures invariantes des Y-systèmes," in *Actes, Congrès intern. Math., Nice, 1970*, Vol. **2** pp. 929–940. Gauthier-Villars, Paris, 1971.

"Gibbsian measures in ergodic theory," *Uspehi Mat. Nauk* **27**, No. 4, 21–64 (1972). English translation, *Russian Math. Surveys* **27**, No. 4, 21–69 (1972).

S. Smale, "Differentiable dynamical systems," *Bull. Amer. Math. Soc.* **73**, 747–817 (1967).

M. Smorodinsky, *Ergodic Theory, Entropy.* Lecture Notes in Mathematics No. 214. Springer, Berlin, 1971.

W. G. Sullivan, "Potentials for almost Markovian random fields," *Commun. Math. Phys.* **33**, 61–74 (1973).

G. Velo and A. S. Wightman (eds.), *Constructive Quantum Field Theory*. Lecture Notes in Physics No. 25. Springer, Berlin, 1973.

P. Walters, "A variational principle for the pressure of continuous transformations," *Amer. J. Math.* **97**, 937–971 (1976).

Ergodic Theory. Introductory Lectures. Lecture Notes in Mathematics No. 458. Springer, Berlin, 1975.

"A generalized Ruelle Perron–Frobenius theorem and some applications," Astérisque **40**, 183–192 (1976).

"Invariant measures and equilibrium states for some mappings which expand distances," *Trans. Amer. Math. Soc.*, to appear.

R. F. Williams, "Classification of subshifts of finite type," *Ann. of Math.* **98**, 120–153 (1973). Errata, *Ann. of Math.* **99**, 380–381 (1974).

Index

Algebra at infinity, 19
Anosov diffeomorphism, 143
Anosov's closing lemma, 124
Axiom, *see* Condition
Axiom A diffeomorphism, 143

Basic set, 124
Bernoulli flow, 163
Bernoulli shift, 79, 128, 155
Borel set, 154
Boundary term, 16

Cluster property, 19, 21, 40
  exponential, 96
"Coarse-grained" quantity, 49
Cohomological interpretation, 161
Commuting continuous maps, 112
Condition
  (A), (B), (C), 20
  (A′), (B′), (C′), 21
  (D), 57
  (D*), 22
  (E), 138
  (M), 149
  (M1), (M2), (M3), (M4), 24
  (M5), 62
  (SS1), 121
  (SS2), 122
  (SS3), 128, 161
  (T), (T′), (T″), 149
  (+T), (+T′), (+T″), 148
Conditional probabilities, 28, 160
Configuration space, 3, 7, 11
Conjugate points, 131
Conjugating homeomorphism, 131
Contour, 31
Cover, 102

DLR equations, 23
Dynamical system
  abstract, 154
  topological, 148
  topologically mixing, 149
  topologically transitive, 149
  topologically + transitive, 148

Energy, 4, 12
Entropy
  on finite set, 3, 42
  mean, or measure theoretic, 6, 43, 103, 155, 162
  topological, 5, 115
Equilibrium state, 6, 40, 110, 126, 163
Ergodic decomposition, 39, 49, 101, 159
Ergodic states, *see* States, ergodic
Expanding map, 138
Expansive $\mathbf{Z}^{\nu}$-action, 4, 101
Extremal point, 151

Face of compact set, 157
Flow, 162
  Smale, 161
  special, 163
Fluctuations, 49
Function A_{Φ}, 34
Function $\psi_{>}$, 81

Generic property, 146
Gibbs ensemble, 3, 13
Gibbs phase rule, 49

Gibbs state, 7, 14, 60, 133, 161
  pure, 20
  unique, 21

Hard cores, 12, 68

Interactions, 12
  exponentially decreasing, 83
  finite range, 34
  invariant, 34
  pair, 55
  physically equivalent, 61
  space $\mathfrak{A}$, 34
  space $\mathfrak{A}_0$, 34
  space $\mathfrak{B}$, 57
  space $\mathfrak{B}_1$, 70
  space $\mathfrak{B}^\theta$, 83
Isomorphism, 25
  $\mathbf{Z}^\nu$-isomorphism, 62

Kolmogorov-Sinai invariant, 103, 155

Lattice (=Crystal lattice), 7, 11, 34
Lattice (=ordered set), 146
Lattice gas, 11, 55, 68
Lattice system, 24
  mixing, 70, 78
  restriction to a subgroup, 64
  sum and product, 32
  transitive, 69
  $\mathbf{Z}^\nu$-lattice system, 62
Lebesgue number, 102
Lebesgue space, 154
Limit
  thermodynamic, 13, 14
  Van Hove, 43
Linear functional
  P-bounded, 152
  tangent, 152
Local product structure, 121

Markov partition, 124, 140
Measure
  abstract, 154
  Bowen, 132
  maximal, 158
  probability, 153
  see also State

Measureable set, 154
Mixture, 49

Morphism, 24
  $\mathbf{Z}^\nu$-morphism, 62

Nearest neighbors, 63
Non-periodicity, 65
Non-wandering point, 148
Non-wandering set, 124, 148

Observable, 12
Operator $\mathfrak{L}$, 80
Operator $\mathfrak{M}_A$, 20
Operator $\mathfrak{S}$, 82

Partition, 102, 155
  generating, 155
  ε-independent, 155
  weak Bernoulli, 156
Partition function, 3, 35, 104, 133
  with periodic boundary condition, 90
Phase
  coexistence, 53
  mixture, 49
  pure thermodynamic, 8, 49
  rule, 49
  transition, 8, 160
Pressure, 4, 37, 104, 126, 163
  relative, 115
  strict convexity of, 61
Property, *see* Condition
Pseudo-orbit, 123

Rectangle, 125
Residual set, 146
Resultant of measure, 157

Separated set, 105, 162
Set $\Lambda(a)$, 36
Shadowing, 123
Shub's entropy conjecture, 161
Simplex, 158
Smale space, 121
Space $\mathcal{C}^\alpha$, 123, 126
Space $\mathcal{F}^\theta$, $\mathcal{F}^\theta_>$, 84
Spaces F^θ, $F^\theta_>$, 85
Space of interactions, *see* Interactions, space
Spanning set, 106
Spectral decomposition, 124
State, 12, 101
  ergodic, 39, 101, 159
  invariant, 39
  see also Equilibrium state, Gibbs state

Subadditivity, 147
  strong, 42
Subcover, 102
Subshift of finite type, 69
Symbol set, 125
Symbolic dynamics, 125
Symmetry breakdown, 8, 59
Symmetry group, 33

Temperature, 4
Theorem
  Abramov, 163
  Bishop-Phelps, 152
  central limit, 129
  Dobrushin, 160
  Friedman-Ornstein, 156
  Hahn-Banach, 150
  Israel, 50, 112, 152
  Krein-Milman, 151
  Lanford-Robinson, 152
  Markov-Kakutani, 151
  Mazur, 152
  Milman, 23, 152
  Ornstein, 156
  Parry, 133
  Perron-Frobenius, 144
  separation, 151
Thermal equilibrium, 4
Thermodynamic formalism, 3
Transfer matrix, 80
Transition matrix, 125
Translation
  invariance, 33
  of lattice, 34

Undecidability, 65
Upper semi-continuity, 147

Vague topology, 153
Variational principle, 3, 6, 45, 108, 117, 163

Wandering point, 148
Weak topology, 151

Zeta function, 89, 135, 164

Printed in the United States
By Bookmasters